M & E HANDBOOKS

M & E Handbooks are recommended reading for examination syllabuses all over the world. Because each Handbook covers its subject clearly and concisely books in the series form a vital part of many college, university, school and home study courses.

Handbooks contain detailed information stripped of unnecessary padding, making each title a comprehensive self-tuition course. These are amplified with numerous self-testing questions in the form of Progress Tests at the end of each chapter, each text-referenced for easy checking. Every Handbook closes with an appendix which advises on examination technique. For all these reasons, Handbooks are ideal for pre-examination revision.

The handy pocket-book size and competitive price make Handbooks the perfect choice for anyone who wants to grasp the essentials of a subject quickly and easily.

For Robert and Thomas

THE M. & E. HANDBOOK SERIES

BASIC BOTANY

CLAIRE SKELLERN, Ph.D.

Formerly lecturer (part-time) at:
Huddersfield Technical College;
Wakefield Technical College; and
Percival Whitley College of Further Education, Halifax.

and

PAUL ROGERS, Ph.D.

Senior Lecturer,
Division of Life Sciences,
Huddersfield Polytechnic.

MACDONALD AND EVANS

MACDONALD & EVANS LTD
Estover, Plymouth PL6 7PZ

First published 1977
Reprinted in this format 1979

ISBN: 0 7121 0255 8

Printed in Great Britain by
Richard Clay (The Chaucer Press) Ltd
Bungay Suffolk

PREFACE

This HANDBOOK is intended to provide an introduction to the study of plants. It should also serve as a companion volume to two books by P. T. Marshall, already published in the HANDBOOK series, *Basic Biology* and *Biology, Advanced Level.* The book follows the standard method for presenting information in the HANDBOOK series and short Progress Tests and examples of G.C.E. questions have been included at the end of each chapter.

Although covering the major areas of botany, emphasis is placed on plant forms and applied botany and the book should be a useful aid to revision of these aspects of botany for G.C.E. A level Botany and Biology examinations, as well as providing a general revision text for O level or equivalent examinations in Botany. It is not intended as a fully comprehensive revision text for A level studies; the study of plant physiology and metabolism, for example, are given a brief treatment as greater detail may be found in *Biology, Advanced Level,* referred to above.

The book should serve as a useful introduction to the study of plants for students with little previous knowledge. It should prove useful to students taking up the subject for the first time late in their academic careers and should provide background information for students of agriculture and horticulture.

Considerable emphasis has been placed on the importance of applied aspects of botany. Three chapters consider crop botany, problems of world food supplies and applied plant ecology, and reference is made to economic and applied aspects of Botany throughout the text. We hope this will assist students in their approach to some of the broader questions frequently found in G.C.E. examinations in Biology and Botany, especially at Advanced level, as well as in the more recently established syllabuses in Environmental Studies and Social Biology.

Acknowledgments

We would like to thank the Joint Matriculation Board, the Associated Examination Board, the University of London University Entrance and School Examinations Council and the Oxford and Cambridge Schools Examination Board for permission to use the examination questions found in Appendix III.

We would also like to thank Mr. Peter Redmond for his help

and encouragement during the preparation of this book, and David Symes and other members of the editorial staff of Macdonald and Evans for their assistance.

January, 1977

C.S.
P.R.

CONTENTS

LIST OF ILLUSTRATIONS

CHAPTER I

THE PLANT CELL

INTRODUCTION TO THE STUDY OF PLANTS

1. The nature of plants. Plants are living organisms and therefore they normally exhibit most, if not all, of the characteristics of living organisms. These include the following.

(*a*) *An ability to grow and reproduce.* Most plants start life as small immature structures which then increase in size and reproduce when they reach maturity. Plants show a wide range of types of reproduction from simple *vegetative* methods through slightly more complex *asexual* methods to highly specialised *sexual reproduction.*

(*b*) *Irritability.* In biological terms, this means that an organism can detect changes in its surroundings and can respond to them, especially if they present a danger. Plants, like animals, can detect and respond to environmental changes, although they do not do so quite so quickly. Examples are the growth of shoots towards light and roots towards soil. They can detect the seasons of the year, growing when conditions are favourable and adopting a dormant state during adverse conditions.

(*c*) *Movement.* Some of the simpler plants can move freely but most of the movement of more advanced plants is restricted to that of growth. The main reason that higher plants do not move is that they live in an environment which provides all their nutritive requirements. Few plants will exhaust the plot of ground they occupy during their lifetime of minerals and none will deplete the atmosphere of carbon dioxide. Unlike most animals, therefore, they do not have to move to find food.

(*d*) *Nutrition.* All living organisms require food which they use to provide the raw materials for energy, growth and reproduction. Plants require simpler foods than animals. They use simple chemicals and the energy of sunlight to synthesise more complex foods by a process known as *photo-*

synthesis. These synthesised foods can then be stored by the plant for use at a later time.

(*e*) *Respiration*. This is the process by which energy is released from the stored food for use by the plant.

(*f*) *Excretion*. This is the removal of waste materials from the plant. Gases are released into the atmosphere through small pores or openings in the leaves, the bark and the roots Other more solid substances are passed into dying leaves which are then shed, or else they accumulate in the dead cells in the centre of the stem. They can also accumulate in bark.

2. The importance of plants. The saying "all flesh is grass" means that all living organisms ultimately depend on green plants for their food. This is because green plants are able to *make* food from simple chemicals using the energy of sunlight. They are therefore known as *primary producers*, with all living organisms depending, directly or indirectly, on them.

The osprey, for example, is a bird of prey feeding on fish, and will catch pike. A pike, in turn, will feed on many smaller fish including perch, and even the perch will feed on still smaller fish such as the stickleback. In each case, an animal eats a smaller animal, and even the stickleback will eat minute water fleas. But the ultimate source of food in this *food chain* is the food of the water flea, which consists of microscopic plants, the single-celled green algae found so commonly in water. Ultimately, the osprey depends on the algae for its food supply the algae making food by photosynthesis. Even the micro-organisms which decompose a dead osprey are really dependent on the algae.

Man, like other animals, is also dependent on plants for his food, but his dependence stretches far beyond that. Plants provide timber, rubber, cotton, rope, beverages, herbs, scents spices, beer, wine and many other products. They even provide the main sources of energy, for coal, oil and natural gas have all been formed from the decaying remains of plants living in prehistoric eras.

THE PLANT CELL

3. Introduction. Plants, like almost all living things, are made up of units called *cells*. There are many different types of cells which fulfil different functions for plants. Cells vary in size from 0·5 μm to 25–100 μm (μm = micrometre = 1/1 000 millimetre).

The smallness of size of cells means that they are studied by means of microscopes, the light or optical microscope for magnifications up to a thousand (× 1 000), and the electron microscope for higher magnifications (e.g. × 100 000). The plant material has to be cut into thin sections and stained or shaded before it can be studied under either microscope.

4. The plant cell under the light microscope. A generalised plant cell is illustrated in Fig. 1. All cells in a plant show some if not all of the structures indicated in Fig. 1 at some stage in their development.

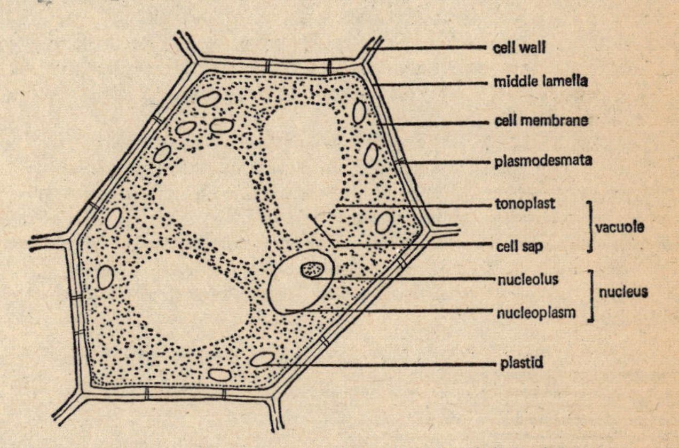

FIG. 1.—A generalised plant cell under the light microscope.

5. The cell wall and middle lamella. The cell wall of a plant is actually made up of two components.

(*a*) *The middle lamella.* This forms a common boundary between cells and is the first part of the cell wall to be formed.

(*b*) *The cell wall.* Composed mainly of *cellulose*, the cell wall forms a semi-rigid case around the cell. It may be made up of two parts, a *primary wall*, that part laid down whilst the cell is still growing, and a *secondary wall*, laid down after growth is complete.

6. Protoplasm. This includes all the contents of the cell within the wall consisting of both the *cytoplasm* and the *nucleus*.

(*a*) *Cytoplasm*. This is a jelly-like substance surrounded by and ramified by small structures, the cell *organelles*.

(*i*) *Cell membrane or plasmalemma*. The cytoplasm is not in direct contact with the cell wall but is surrounded by a thin skin, the *cell membrane*.

(*ii*) *Plasmodesmata*. These are strands of cytoplasm which pass through small pores in the cell wall from one cell to another and are surrounded by a sheath of cell membrane.

(*iii*) *Vacuoles and tonoplast*. The cytoplasm often does not fill the centre of the cell; it may cling to the periphery with strands passing across the middle. The spaces between these strands and smaller spaces in the cytoplasm are called *vacuoles*. Again the cytoplasm is protected by a skin which lines the vacuoles and this is called the *tonoplast*.

(*iv*) *Cell sap*. This is a solution containing dissolved salts which fills the vacuoles.

(*v*) *Plastids*. These are ovoid bodies set in the cytoplasm. Some are green in colour, containing *chlorophyll*, the green pigment which is one of the essentials for *photosynthesis* (the mechanism by which plants manufacture food). These green plastids are *chloroplasts*. Other plastids may be coloured *chromoplasts* found in coloured leaves and petals. Others contain starch grains as a food reserve and are called *amyloplasts*.

(*b*) *Nucleus*. Observed as a round or ovoid body which often stains darkly in microscopic preparations, it is made up of nucleoplasm and organelles.

(*i*) *Nuclear membrane*. This is similar to the cell membrane but it has large pores in it.

(*ii*) *Nucleolus*. Often there is only one nucleolus but occasionally more are seen as darkly staining areas within the nuclear material.

(*iii*) *Chromosomes*. Also set in the nucleoplasm are long thin threads of chromosomes which carry the information for the working of the cell and for inheritance. They are not shown in Fig. 1 because they can only be seen under the light microscope during cell division (*see* **9** below).

PLANT CELL ULTRASTRUCTURE

7. Introduction. If cells are observed at higher magnifications under an electron microscope, much more detail of the structure of the walls, membranes and organelles can be seen. This is the ultrastructure and is represented in Fig. 2.

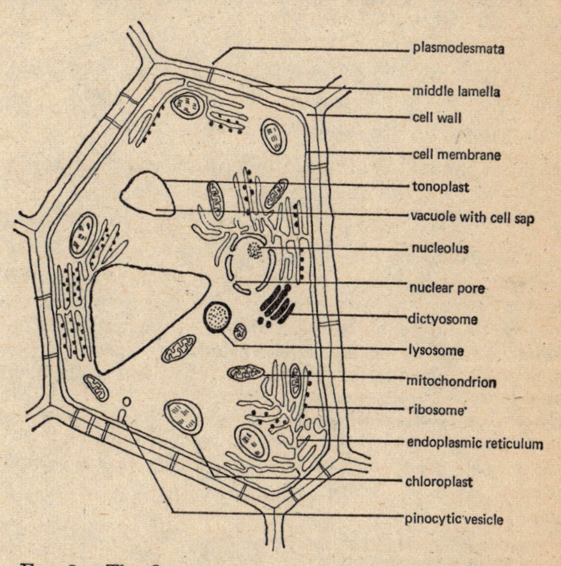

FIG. 2.—The fine structure of a generalised plant cell.

8. The middle lamella. This is made up of a substance called *pectin* which is highly plastic but has no apparent structure. It allows for growth of the cell.

9. The cell wall.

(*a*) *The primary cell wall* is composed of long strands of cellulose called *microfibrils* which are laid down haphazardly within a matrix of hemicellulose and pectin. This also allows the cell to grow and is permeable to water, solutes, salts and sugars.

(*b*) *The secondary cell wall* is also composed of cellulose microfibrils but these are highly organised. The microfibrils are laid down parallel with and close to one another. The matrix consists entirely of hemicellulose and there is much less matrix than in the primary cell wall. The microfibrils are laid down in layers so that they lie at different angles in each layer. Other substances such as *lignin*, *suberin*, *cutin* and waxes may be laid down on the secondary wall or may impregnate it.

10. Protoplasm. This comprises all the material and organelles within the cell wall and is composed of the cytoplasm and organelles and the nucleus and organelles.

(*a*) *Cytoplasm.* Proteins, lipids, nucleic acids and water-soluble substances are found in the cytoplasm. It has a jelly-like or *colloidal* structure in which the organelles are set.

(*i*) *Cell membrane or plasmalemma.* Under the electron microscope this shows up as two dark lines separated by a lighter zone. The two dark lines are protein and the pale zone is found to be lipid. Further investigation has shown that the lipid molecules in the middle are in two layers, and each molecule has a lipophyllic end and a hydrophyllic

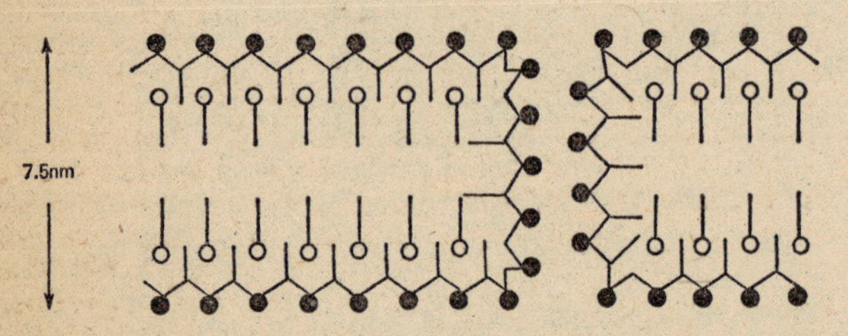

FIG. 3.—A cell membrane.

end. Lipophyllic ends are attracted to lipids and hydrophyllic ends are attracted to water and to water-soluble substances like protein. The lipophyllic ends are attracted to each other and the hydrophyllic ends are attracted to the protein, as shown in Fig. 3. The protein layers are each 2 nm wide (nm = nanometre = 1/1000 μm) and the double lipid layer in the middle is 3·5 nm wide, this making up 7·5 nm for the whole structure which is called a *unit membrane.* The membrane is important in controlling the passage of materials into and out of the cell. Lipid soluble molecules pass easily between the lipid molecules but many other non-lipids also enter or leave the cell easily and it is thought that this takes place through pores in the membrane. These pores are breaks in the lipid layers where the outer and inner protein layers join up so that the pore is lined with protein molecules.

(*ii*) *Pinocytic vesicles.* These are membrane-wrapped parcels which are passed into or out of the cell by *pinocytosis.* This is a process by which substances which are entering the cell are surrounded by a layer of the cell membrane. As the substance is pushed further into the cell, the cell membrane completely encloses it and breaks away from the rest of the membrane. The parcel or pinocytic vesicle is then free to move through the cytoplasm. Substances to be removed are surrounded by a membrane in the cytoplasm and passed to the cell membrane. Here the two membranes fuse and the material is expelled.

(*iii*) *Plasmodesmata.* The cell membrane of one cell is thought to be in connection with the cell membrane of the next cell through pores in the cell wall. There is some controversy about whether the *endoplasmic reticulum* of the two cells is also connected. It is suggested that plasmodesmata may provide an easy route for large molecules to pass from cell to cell.

(*iv*) *The endoplasmic reticulum.* As its name implies, this is a network of membranes running within the cytoplasm. The endoplasmic reticulum (ER) is made up of a pair of unit membranes which are folded and refolded to occupy most of the cytoplasm. The spaces between the unit membranes of each pair are filled with a watery fluid. The area between the folds is occupied by the cytoplasm itself.

(*v*) *Ribosomes.* Attached to the sides of the membranes adjacent to the cytoplasm are small sub-spherical bodies called ribosomes. These are about 20 nm in diameter and have a groove around them into which a template of *ribose nucleic acid* (RNA) fits during protein synthesis (*see* VIII, 6). Ribosomes are often found in chains on the endoplasmic reticulum and these are called *polysomes.* An ER with ribosomes attached is known as a *rough membrane.*

(*vi*) *Mitochondria.* A mitochondrion may be either spherical or sausage-shaped and has dimensions of 1·0 μm in diameter and up to 2·5 μm in length. There may be up to 1 000 mitochondria in a single cell. As can be seen diagrammatically in Fig. 4, they are surrounded by two unit membranes, an outer tight membrane and an inner folded one. The space between the two is filled with liquid. The inner membrane is ridged to increase the surface area. These ridges are called *cristae* and are the sites of respiration (*see* VII, 34). The centre is filled by a matrix.

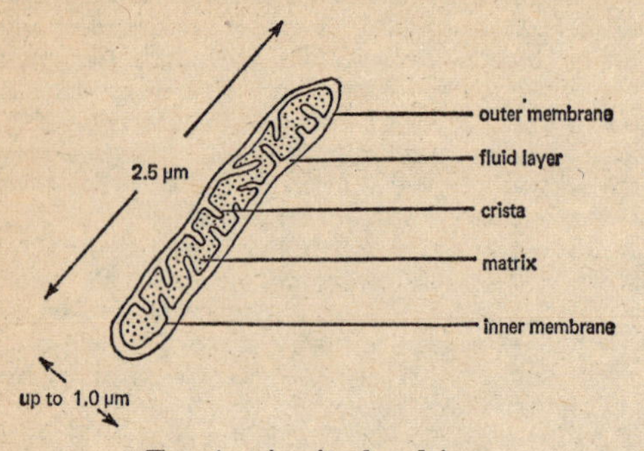

FIG. 4.—A mitochondrion.

(*vii*) *Dictyosome.* Occasionally, even in plants, the dictyosome is called the *Golgi body*. There are a number of dictyosomes in each cell and they are seen as stacks of flattened smooth membrane sacs which are 1 to 5 μm in diameter and 0·5 μm thick (*see* Fig. 5). The edges of the sacs

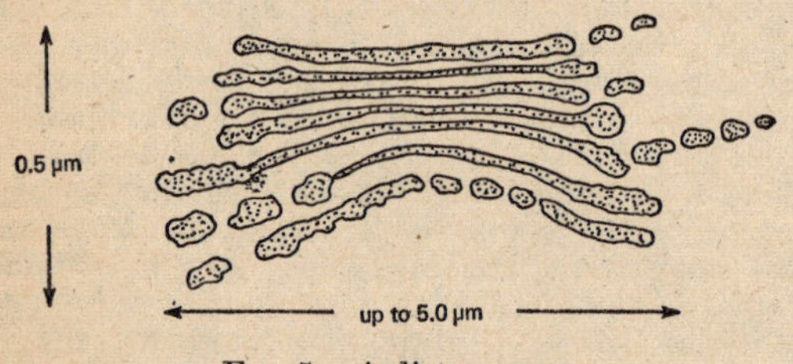

FIG. 5.—A dictyosome.

are thicker and from these edges bits are pinched off to produce separate little membrane-surrounded spheres which can move freely in the cytoplasm. It is thought that various polysaccharides are secreted and transported in this manner. They may even be expelled from the cell by pinocytosis.

(*viii*) *Lysosome.* This is a membraneous sac which is filled with enzymes capable of digesting the cell if the sac is ruptured. It is often called the suicide bag. A badly damaged cell can digest itself and the surrounding healthy cells.

can then take up the simple constituents left over after this process.

(*ix*) *Vacuoles and tonoplast.* There is little more to say of these when viewed under the electron microscope except that the tonoplast, like the cell membrane, is a unit membrane.

(*x*) *Chloroplasts and plastids.* Chloroplasts are disc-shaped bodies up to 5 μm in overall diameter. They are surrounded by two unit membranes as can be seen in Fig. 6.

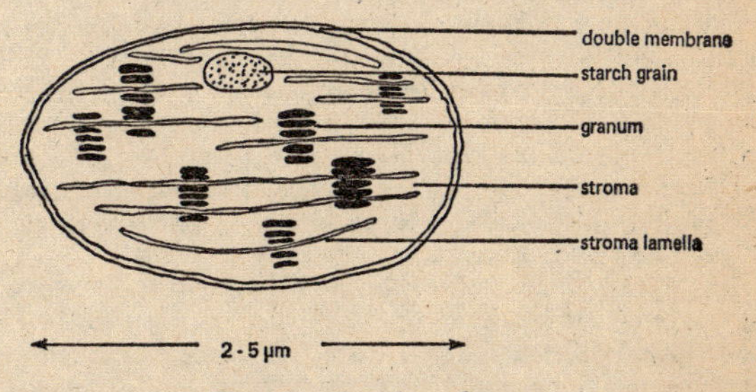

FIG. 6.—A chloroplast.

The interior is also made up of layers of membranes passing through a matrix which is called the *stroma.* The membrane pairs running in the stroma are called *stroma lamellae* and are close together. In certain areas, called the *grana* (singular *granum*), the membranes are pushed slightly apart by short lengths of another membrane being introduced between them. On the surfaces of the membranes of the granum are raised areas called *quantosomes,* each of which contains an estimated 230 molecules of chlorophyll. Starch grains are also found in the stroma. The quantosomes are the sites of photosynthesis in the cell (*see* VII, 25), and starch is an end product of photosynthesis, hence its location as granules in chloroplasts. Other plastids have very little structure. The most common, *amyloplasts,* are made up of a number of starch grains surrounded by a membrane.

(*b*) *Nucleus.* Like the cytoplasm, the nucleus is surrounded by a membrane. It also has a jelly-like matrix, the nucleoplasm, in which are to be found various organelles.

(*i*) *Nuclear membrane.* This, like the endoplasmic reticulum, is a double membrane. Although it does not have ribosomes on the inner side, it may have them on the cytoplasm side, so that the inner side is always smooth. It forms the boundary of the nucleus.

(*ii*) *Nucleopore.* These are pores in the nuclear membrane where the inner membrane joins the outer one and are thought to provide a means whereby transport can take place between the nucleus and cytoplasm.

(*iii*) *Nucleoplasm.* The centre of the nucleus is filled with a material that has a similar consistency to cytoplasm but has different constituents. In the nucleoplasm are found areas of *chromatin* and *chromosomes.*

(*iv*) *Nucleolus.* Sited in the nucleoplasm is at least one nucleolus and occasionally two. They are darkly staining bodies whose precise function so far remains unknown, although they contain large amounts of ribose nucleic acid and protein and are involved in nucleic acid production.

(*v*) *Chromatin.* This forms darkly staining areas in the nucleoplasm and is also the material of which chromosomes are made. It consists of large amounts of *deoxyribose nucleic acid* (DNA) and protein.

(*vi*) *Chromosomes* are seen at all times under the electron microscope as dark long thin double threads. The two threads are chromatids and are held together at one narrow point somewhere along their length which is known as the *centromere.* Chromosomes are made up of units called *genes* which provide the information for running the cell and which pass information on from one generation to the next, i.e. inheritance. When chromosomes are seen under the electron microscope they are seen to have light and dark bands across them but so far no relationship has been found between these bands and genes. There are a number of chromosomes in a nucleus and these differ in their length, shape and the position of the centromere. In most plant cells there is at least a pair of each type in the nucleus. The exception is found in the reproductive cells where there is only one of each type so that on fertilisation the pairs are restored.

CELL DIVISION

11. Mitosis. Plants, like all living things, grow. They do this by cell division, the daughter cells then enlarging. The process by which cells in the growth or *meristematic* areas of the plant divide is somatic division or *mitosis.* The areas where mitosis is normally found are the meristems in the shoot and root tips.

Mitosis is really a continuous process but it is divided into stages for convenience. Cells have a resting stage or *interphase* between each division and the chromosomes cannot be seen under the light microscope during this stage.

At the onset of mitosis or *prophase* the chromatin which was scattered throughout the nucleus appears to coalesce into coiled chromosomes. The two chromatids and the centromere can also be distinguished.

The next thing to happen is that the nuclear membrane ruptures and the nucleolus disappears. This is termed the *prometaphase*. Also during prometaphase a spindle appears consisting of a group of fibres running from a point or pole at one side of the cell, widening to a maximum near the centre and then narrowing to another pole on the other side. Each chromosome becomes attached to a spindle fibre by its centromere. The chromosomes then move along the spindle fibres with the centromeres leading them.

During the next stage, *metaphase*, the chromosomes are found attached to the spindle by their centromeres with the chromatids lying free outside the spindle area. They lie at the widest part or equator of the spindle where they form the *metaphase plate*.

Anaphase begins with the splitting of the centromere and the movement of the chromatids to opposite ends of the spindle. The centromere leads the movement and the chromatids or *daughter chromosomes*, as they are now called, trail behind. In the latter part of anaphase and at the beginning of *telophase*, a layer of nodules or *phragmoplasts* is formed between the two groups of separating daughter chromosomes. Also during telophase a nuclear membrane is formed around each group of daughter chromosomes which gradually fade to be replaced by chromatin patches when seen under the light microscope. The nucleolus is also re-formed so that the nuclei then take on the interphase state.

During the late anaphase and telophase, a *cell plate* forms in the phragmoplasts. This cell plate gradually extends to the edges of the cell and long after the nuclei have returned to the interphase state it reaches the cell wall and divides the old cell into two new cells. The two new cells then enter an interphase in which the cells enlarge a little. The daughter chromosomes must also each form another chromatid because at the beginning of the next prophase they can be seen to have two chromatids. In onion root tips the actual division takes about 4 hours and then there is a rest period of from 14 to 20 hours so that the whole cycle takes from 18 to 24 hours. Fig. 7 shows the process of mitosis.

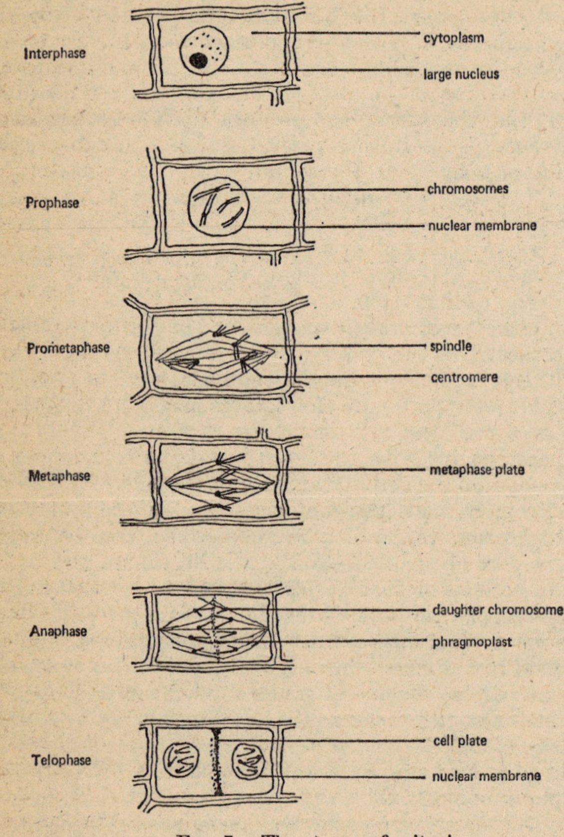

FIG. 7.—The stages of mitosis.

12. Meiosis. An ordinary plant is made up of cells which contain two sets of chromosomes, one derived from the male parent and one from the female parent. These plants are said to be *diploid*. If, however, all these chromosomes passed into the *gametes* (the reproductive cells) after fertilisation, there would be a doubling of the number of chromosomes in the next generation and a quadrupling in the generation after that. As this does not normally happen, there has to be a mechanism whereby only one set of chromosomes is passed into the gametes. This set of chromosomes is made up of one of each type of chromosome which may have come from either parent. The mechanism which reduces the chromosome number to half is *meiosis* or *reduction division*. Cells with only one set of chromosomes are said to be *haploid* or sometimes *monoploid*.

Meiosis consists of two stages:

(*a*) *the first stage* is a reduction division which produces two haploid nuclei;

(*b*) *the second stage* duplicates each of these nuclei and the result is therefore four haploid nuclei.

These two stages used to be regarded as separate divisions called respectively the heterotypic and homotypic division. They are now regarded as stages of meiosis and are referred to as the *first meiotic* and the *second meiotic division*. Meiosis, like mitosis, is divided up into phases for convenience but it is a continuous process.

During *prophase* I of the first meiotic division, the nucleolus disappears and the chromosomes become shorter and thicker and are seen to consist of two chromatids joined by a centromere.

Like chromosomes then begin to move towards one another, a process called synaptic pairing of homologous chromosomes; this results in similar chromosomes lying next to one another in pairs called *bivalents*.

The chromatids of the two chromosomes then become intertwined and where they touch one another they form a *chiasma* (plural *chiasmata*) at which the bivalents become attached. At a chiasma the two chromatids break and each part rejoins on to the opposite chromatid forming a *cross-over*.

Towards the end of *prophase* I, a spindle is formed and at *metaphase* I the bivalents move along the spindle to lie on the equator of the spindle. The orientation of a bivalent is random and quite independent of the other bivalents. At this stage, also, the nuclear membrane disappears.

Anaphase begins with the separation of the homologous chromosomes of the bivalents, one moving to one pole of the spindle and the other moving to the other pole. Due to the random nature of the attachment of the bivalents to the spindle at the equator of the cell, the separation of the chromosomes is also random. As the bivalents move apart, the chiasmata are broken (*see* Fig. 8).

Telophase finds the homologous chromosomes with their exchanged segments at different ends of the spindle. It would

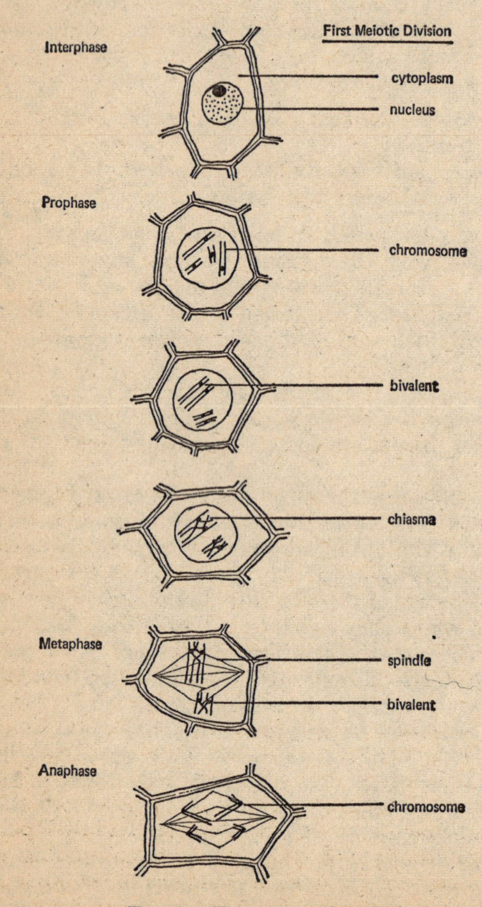

FIG. 8.—The stages of meiosis.

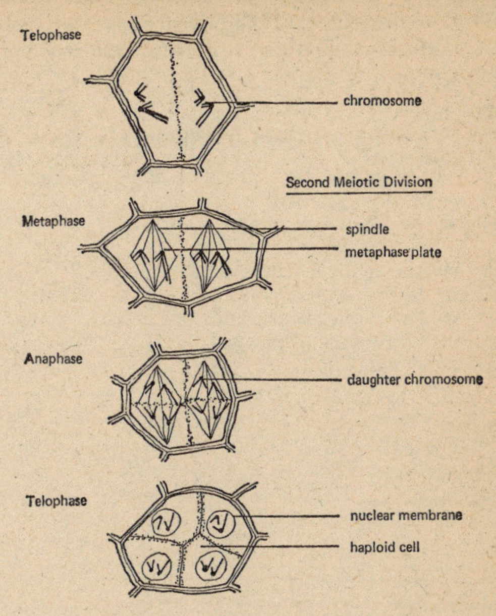

Fig. 8.—The stages of meiosis (contd).

now be expected that two separate cells would be formed and after a resting period or *interphase* the second meiotic division would begin, but in fact plant cells move directly from *telophase* I to *metaphase* II. This is achieved by the cell producing two new spindles at right angles to the first one, with their equators where the poles of the original spindle were, so that the chromosomes now form a *metaphase plate*. At *anaphase*, the chromatids of each chromosome separate to go to opposite poles of the spindles.

During *telophase* II, the new cell walls are formed, the spindles disappear and the nuclear membrane and nucleolus reappear. This results in four haploid cells commonly referred to as a *tetrad*. These haploid cells may act as gametes which require fusion before further development can take place, as in the Thallophyta, or they may germinate to produce prothalli which then produce gametes, as happens in the higher plants.

The exchange of genetic material in crossing over and the

independent segregation of the chromosomes ensures that new genetic combinations can arise and this is a basic factor in evolution.

PROGRESS TEST 1

1. What are the main characteristics of living organisms? **(1)**
2. Why are plants important? **(2)**
3. Describe the main features of a plant cell as seen under the light microscope. **(3–6)**
4. What are the main features of the plant cell wall? **(9)**
5. What are mitochondria and chloroplasts? **(10)**
6. What are the main features of the nucleus? **(10)**
7. Describe the process of mitosis. **(11)**
8. What is meiosis? **(12)**

CHAPTER II

VIRUSES, BACTERIA AND FUNGI

1. Introduction to plant classification. Plants are divided into groups depending on the complexity of organisation within the plant body and on the type of reproduction they exhibit.

The simplest structures described here are the viruses; they are not typical of living organisms and are often left out of plant and animal studies. Apart from viruses, the plant kingdom is divided as follows.

CRYPTOGRAMS (flowerless plants)

Thallophyta: including bacteria, fungi, lichens and algae; they have a simple structure to the body and little or no differentiation.

Bryophyta: including mosses and liverworts; they show some differentiation into leaf and rhizoids and have vascular tissue.

Pteridophyta: these include ferns and have differentiation into stems, leaves and roots with clearly defined vascular tissue.

PHANEROGRAMS (flowering plants)

Gymnosperms: these include conifers and have naked seeds.

Angiosperms: these include monocotyledons such as grasses and dicotyledons such as roses. The seed is enclosed in an ovary.

VIRUSES

2. Habitat. Viruses are parasitic organisms found in plants, animals and bacteria.

3. Size, shape and structure. Viruses are very small particles which pass through porcelain filters designed to retain bacteria. They are between 0·012 μm (12 nm) and 0·8 μm (800 nm) in size. Because of their small size, viruses are studied under the electron microscope. This has shown them to consist of many shapes including spheres, flexible and rigid rods, ellipses and polyhedrons. The bacterial virus or *bacteriophage* has a polyhedral head with a tail and an end plate.

4. Structure. All viruses are made up of an outer layer of protein composed of distinct sub-units and a core of nucleic acid (*see* VII, **11**). This is DNA in bacteriophages and some animal viruses or RNA in most animal and all higher plant viruses. The protein provides a protective sheath for the inheritable and infective material, the nucleic acid. Figure 9 shows the structure of a typical bacteriophage virus.

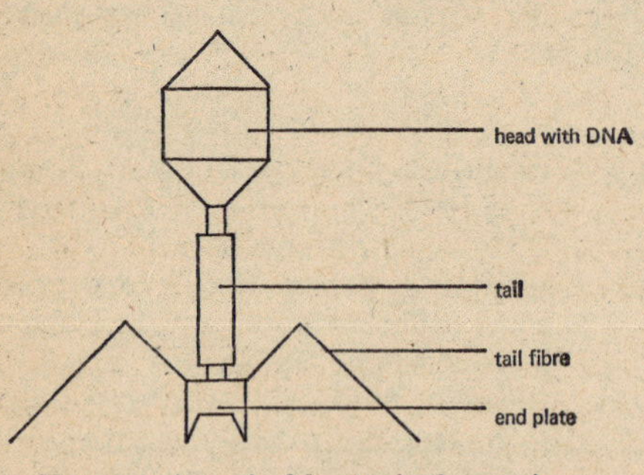

FIG. 9.—A typical bacteriophage virus.

5. Nutrition and reproduction. Virus particles may be transferred from an infected host to a new host by direct contact or with the aid of some external agent or *vector*. For example, aphids transfer a number of plant viruses and mosquitoes transfer some animal viruses. Bacteriophages are injected into the bacterial cell through special receptive areas on the wall.

The protein sheath of the virus is often left on the outside of the cell and the nucleic acid is passed inside to redirect the metabolism of the host cell to make it produce new viral particles. The virus is not only parasitically dependent on the cell for raw materials but it also uses the cell's energy resources. New nucleic acid cores are produced as are new protein sheaths, and then the two are assembled. With a bacteriophage, 300 new viral particles may be produced in a single bacterium which then bursts, liberating the virus.

NOTE: In bacterial viruses the viral nucleic acid may become a *provirus* by attaching itself to the bacterial nucleic acid and reproducing along with the bacterial nucleic acid. This is the *temperate state*. After some time the provirus may take over the cell and become *virulent* once more.

Virus multiplication can thus only take place in a living cell because of the need for an energy source. Virus particles, however, can survive long periods in adverse conditions. Tobacco mosaic virus can survive 50 years in tobacco, 2 weeks in dried sap on clothing and up to 10 minutes boiling.

6. Economic importance. Many diseases of man and other animals are caused by viruses, including poliomyelitis, yellow fever and the common cold in man, and foot and mouth disease and rinderpest in cattle. Almost all crop plants and many wild plants can be affected by viruses and these may produce mottling, stunting, reduction in vigour, distorted growth and even colour-breaks in flower petals. A virus disease has recently been found in mushrooms.

Viruses are not normally affected by antibiotics but there is, however, a protein substance *interferon* which is produced by many animals which does affect viruses. Interferon is non-toxic to the host and non-specific, but unfortunately it has not yet been produced commercially.

BACTERIA (SCHIZOMYCOPHYTA)

7. Habitat. Bacteria are ubiquitous. They may be parasitic on plants and animals, saprophytic on all kinds of food and present in soil and on almost any organic matter. They are also found in water. Except under extremes of temperature it is almost impossible to find a part of the earth's surface where bacteria are not active.

8. Size, shape and structure. Bacteria can be seen with a light microscope when suitably stained and their sizes range from 0·2 to 2·0 μm in diameter and up to 10 μm in length. There are three distinct shapes:

(*a*) *cocci* (sing. *coccus*)—spheres;

(*b*) *bacilli* (sing. *bacillus*)—rods;
(*c*) *spirilla* (sing. *spirillum*)—spirals.

Cocci are usually non-motile but motile forms of the bacilli and spirilla are found. Motility is brought about by *flagellæ*, which are whip-like threads which may occur singly or in large numbers, located terminally, laterally or both.

All bacteria have a rigid cell wall made up of a complex of amino acids, sugars and fats. Some have a loose slime layer around them while others have a very compact layer which forms a *capsule*. This provides protection for the bacterium and the presence or absence of a capsule may determine the effectiveness of an antibiotic.

Within the wall is a plasma membrane which is the site of respiration as there are no mitochondria. The cytoplasm contains ribosomes and the hereditary material DNA is in a nuclear body but this is not a true nucleus as there is no nuclear membrane.

Under adverse conditions, some bacteria, particularly bacilli, can produce *endospores*. The contents of the cell shrink and a thick wall develops around them. These endospores have a low metabolic rate and a low water content. The old cell wall often disintegrates to leave the oval resting endospore. When conditions favourable for growth return, the spore takes up water, its coat ruptures and a normal cell emerges.

9. Nutrition. Many different types of nutrition are found in bacteria, some being *autotrophic* and others *heterotrophic*.

(*a*) *Autotrophic bacteria* are able to synthesise their own carbon compounds from carbon dioxide, water and other chemicals.

(*i*) *Photoautotrophic bacteria*, which are rare, can use light as their source of energy in the process of photosynthesis. They are most common in shallow lakes.

(*ii*) *Chemoautotrophic bacteria* use chemical energy instead of light. Perhaps the most important members of the group are the nitrifying and denitrifying bacteria which are essential components of the nitrogen cycle (*see* IX, **13**).

(*b*) *Heterotrophic bacteria* rely on previously synthesised food sources and are either *saprophytic*, utilising dead organic matter, or *parasitic*, utilising living organic matter and usually causing a disease in the host organism. Bacteria secrete digestive enzymes on to their food source and the soluble nutrients which are thereby released are taken into the bacterium either by *diffusion* or by *active transport* (*see* VII, 22).

10. Oxygen requirements. Some bacteria require atmospheric or free oxygen and are called *aerobes* but others utilise chemical compounds containing oxygen and may even be inhibited by free oxygen. These are called *anaerobes*—fermentation processes are brought about by anaerobes.

11. Heat requirements. The majority of bacteria require a temperature of between 20°C and 46°C although a few live best in cold conditions down to 0°C (some live in refrigerators). Still others grow best at higher temperatures, from 40°C to as high as 75°C with a few bacteria living in hot springs.

12. Toxin production. Small amounts of poisonous materials are produced by some bacteria. These not only limit their own growth but can have serious effects on other organisms. *Claustridium botulinum* produces the toxin responsible for the dangerous food poisoning condition of *botulism.*

13. Reproduction. This is brought about by the inward growth of the cell wall to divide the bacterium into two parts, hence the original name for bacteria, *schizophyta,* meaning fission plants. The cytoplasm and nuclear material is divided between the two parts but not by mitosis. Some bacteria can undergo a division every 20 minutes and it is this ability to reproduce at a phenomenal rate which allows bacteria to take advantage of a suitable environment very quickly. This can be shown as follows:

Time	*No. of bacteria*
0 min	1
20 min	2
40 min	4
1 h 0 min	8
1 h 20 min	16
1 h 40 min	32
2 h 0 min	64
4 h 0 min	4096

The increase in numbers is slow at first (the *lag phase* of growth) followed by a sharp increase in rate of growth (the *log* or *logarithmic phase*) which lasts until the food source is depleted or until waste products start to poison the bacteria.

14. Economic importance. Bacteria are utilised for producing cheese, vinegar, yoghurt, silage and some industrial chemicals. They can, however, spoil food and other materials and cause diseases such as tuberculosis and sepsis in man, anthrax in cattle and many diseases of plants such as blackleg of potatoes and fireblight of pears.

Many diseases of animals and man can now be controlled or reduced by *immunisation*. This involves the introduction of weakened or heat-killed bacteria into the body so that it produces its own *antibodies*. These antibodies are then retained in the blood to counteract the virulent bacteria should they attack the body at a later time.

Antibiotics such as penicillin have been isolated from fungi and actinomycetes and are produced commercially for treating bacterial infections. They are usually very effective at first but bacteria can ultimately become resistant to a particular antibiotic and there is therefore a constant search for new and improved antibiotics.

FUNGI

15. Habitat. Fungi are most commonly saprophytic on many kinds of organic material, but they can also be parasitic on plants and animals. Fungal spores are widely distributed in the air and can readily take advantage of new food sources.

16. Size and shape. Fungi range in size from microscopic yeast cells up to large bracket fungi 0·5 metres across which are found on many trees. The shape of the fungal body varies greatly. While some are single cells, others, although not showing great division of labour internally, have complex structures for spore production, protection and dispersal.

17. Structure. Most fungi are made up of thin transparent threads which contain cytoplasm, nuclei and vacuoles. These threads are called *hyphae* (*sing. hypha*). An intertangled mass of these hyphae is termed the *mycelium*. The hyphal walls are made up of cellulose or chitin and it is this fact along with their lack of motility which puts fungi in the plant kingdom, even though they do not possess chlorophyll.

18. Nutrition. Fungi, because of this lack of chlorophyll, are heterotrophic, living either saprophytically or parasitically. They release enzymes on to their food and subsequently absorb the soluble products through their cell walls.

19. Reproduction. Many fungi can reproduce *asexually*, producing large numbers of genetically identical spores which helps them to colonise a new area quickly. They may also reproduce *sexually* with a few new individuals produced following fusion of two nuclei.

20. Classification. The fungi are divided into three main groups.

(*a*) *Phycomycetes.* The hyphae in phycomycetes have no cross walls, that is they are not divided into cells and are said to be *coenocytic.*

(*b*) *Ascomycetes.* The hyphae have cross walls and the spores produced in sexual reproduction are produced inside sac-like structures called *asci* (sing. *ascus*).

(*c*) *Basidiomycetes.* The hyphae have cross walls but the sexually produced spores are produced on the outside of structures called *basidia* (sing. *basidium*).

PHYCOMYCETES

21. The phycomycete life cycle. A typical life cycle for a phycomycete is shown in Fig. 10.

22. Pythium debaryanum is an example of a phycomycete with this type of life cycle. It is a fungus which causes "damping off" in seedlings of many plants. The fungus is present in soil, attacking the seedlings at ground level, penetrating the tissues and causing the seedling to fall over. Under damp conditions, hyphae are pushed through the infected tissue and swellings called *zoosporangia* are produced at the ends. From each of these, a sac is extended, the contents of the zoosporangium pass into the sac and *zoospores* are produced. These are small round bodies with two whip-like flagellae which allow

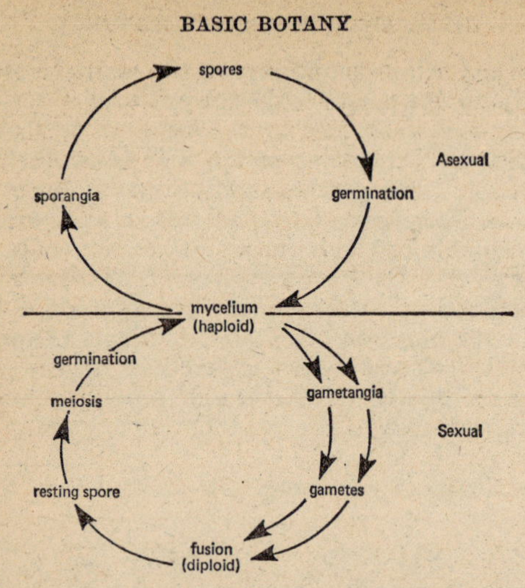

FIG. 10.—A generalised phycomycete life cycle.

them to disperse through surface moisture. The zoospores subsequently lose their flagellae and encyst for a time and eventually germinate to produce a new mycelium capable of infecting a healthy seedling. The mycelium and zoospores contain one set of chromosomes and are *haploid*.

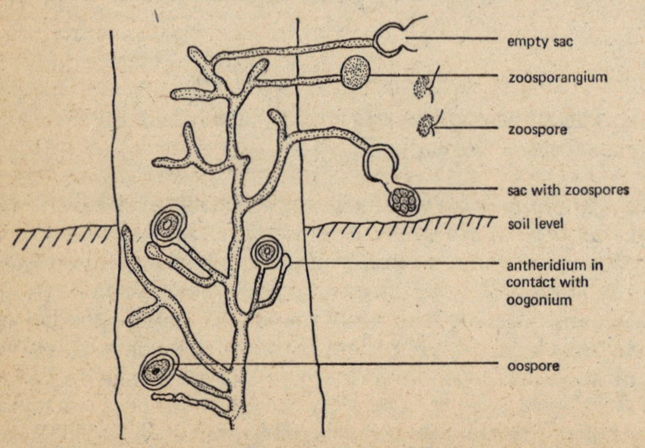

FIG. 11.—The detailed structure of *Pythium debaryanum*.

When conditions are not so favourable and the seedling has died, two adjacent hyphae produce *gametangia*. Cross walls are formed to separate them from the rest of the hyphae and all the nuclei except one in each gametangium degenerate. A thin *antheridium* is produced with a round *oogonium* adjacent to it. The two become connected by a *fertilisation tube*. The nucleus from the antheridium then passes into the oogonium along the fertilisation tube and fuses with the oogonium nucleus to produce an *oospore*. The oospore has two sets of chromosomes, one from each nucleus, and is therefore *diploid*. It develops a thick resistant wall and is only released by the rotting of the seedling tissue. When the oospore finally germinates, it has already undergone *meiosis* or reduction division and the new mycelium is again haploid.

23. Economic importance. Phycomycetes are responsible for many of the mould diseases of fruit and vegetables. They cause other plant diseases such as downy mildews of lettuce and grapes and potato blight, and they also cause diseases of fish.

ASCOMYCETES

24. The ascomycete life cycle. A generalised ascomycete life cycle is shown in Fig. 12.

Ascomycetes have septate hyphae (i.e. with cross walls) which are divided into cells, and they produce sexual spores inside asci.

25. Erysiphe graminis is a typical ascomycete and causes powdery mildew diseases of many grasses and cereals. The mycelium forms whitish-grey cotton-like patches over the leaves of infected plants and the hyphae have special structures called *haustoria* which penetrate the mesophyll cells of the host to draw on its food reserves. When conditions are favourable, aerial hyphae called *conidiophores* grow up and these produce long chains of *conidia* which eventually break off, are dispersed, and germinate to produce a new mycelium, perhaps on another part of the leaf.

26. Sexual reproduction takes place towards the end of the host's growing season. The end of one hypha produces an *antheridium* and another hypha forms an *ascogonium*. Both of these gametangia have only one nucleus each, i.e. they are *uninucleate*. The two come close together and the nucleus from the antheridium passes into the ascogonium. The nuclei do not fuse but divisions of the nuclei and the ascogonium take place to form an *ascogenous hyphae* with *binucleate cells*. These cells of the ascogenous hyphae are known as the *ascus mother*

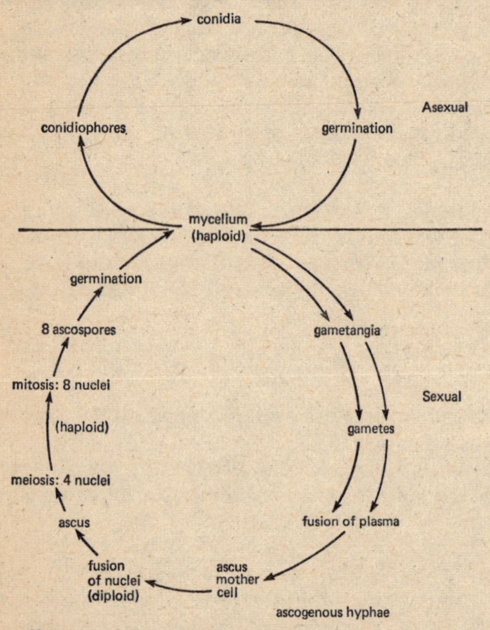

FIG. 12.—A generalised ascomycete life cycle.

cells and between eight and twenty-five of them develop asci which are long tubular outgrowths.

The two nuclei in each ascus mother cell fuse so that they are now diploid. Meiosis then follows producing four haploid nuclei and this is in turn followed by mitosis giving eight haploid nuclei. Occasionally eight, but often only four, of these nuclei develop into the ascopores within the ascus formed by the ascus mother cell. Whilst the asci and ascospores have been developing, hyphae from the surrounding mycelium have

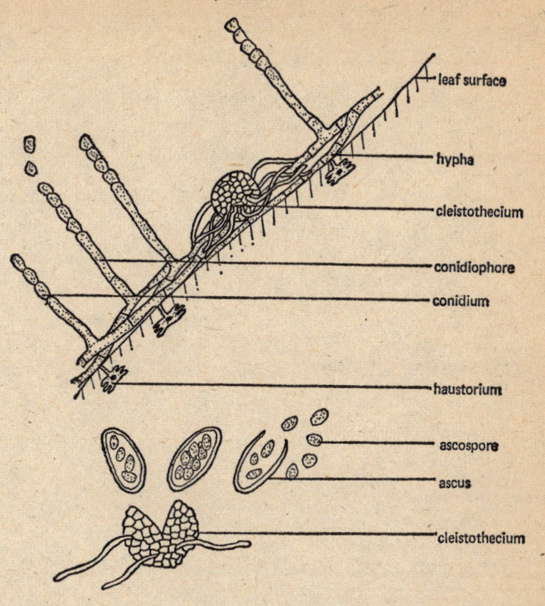

FIG. 13.—The detailed structure of *Erysiphe graminis*.

grown up to provide a protective ball around them called a *cleistothecium*. This eventually bursts liberating eight to twenty-five asci each with four to eight ascospores. The haploid ascospores are then released and disperse to germinate later and form a new mycelium.

27. Economic importance. Atypical but well-known ascomycetes are the yeasts widely used in baking and brewing. These are single-celled fungi which reproduce asexually by budding but which occasionally produce four to eight ascospores. The powdery mildew diseases of apples, gooseberries, roses and peaches are caused by ascomycetes, as is apple scab disease. Green and black moulds on many foods, fruit and vegetables in store are cause by *Penicillium* and *Aspergillus*. Both these ascomycetes produce antibiotics which are widely used in medicine. Truffles and the fungal part of lichens (*see* III, **27**) are also ascomycetes.

BASIDIOMYCETES

28. The basidiomycete life cycle is shown in Fig. 14. The life cycle, as can be seen, is complex, especially with the different forms of asexual reproduction. All the aspects of the life cycle

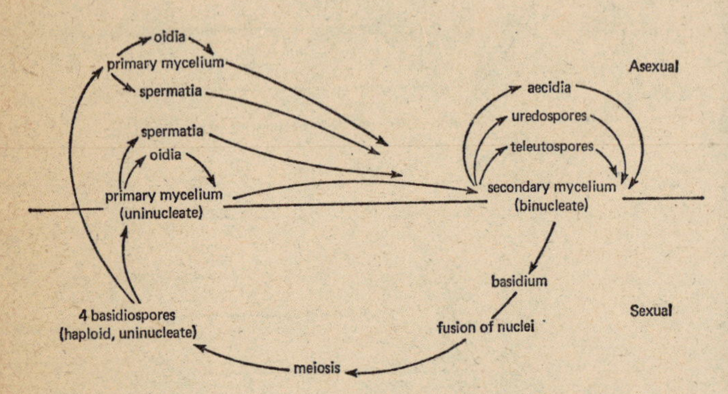

FIG. 14.—A generalised basidiomycete life cycle.

are not necessarily found in any one species of the basidiomycetes, the distinguishing features of the whole group being a septate hypha and the production of sexual spores outside basidia.

29. Agaricus campestris, the common mushroom, is one of the best known basidiomycetes. In this fungus the uninucleate basidiospores germinate to produce the *primary mycelium* which is formed of uninucleate cells. A *secondary mycelium* develops from the meeting of two uninucleate hyphae with the nucleus of one passing into the other to form a binucleate cell.

NOTE: Some basidiomycetes have primary mycelia which produces *oidia* which, in turn, can germinate to produce more primary mycelia. Primary mycelium may also produce *spematia*. These do *not* form further primary mycelium but will unite with a primary mycelium hypha to form a binucleate cell.

When the binucleate cell divides, it produces a secondary mycelium. During this process a special structure called a *lamp connection* is produced to ensure that all new cells are he same as the original binucleate cell. Its two nuclei A and B livide to form A′ A′ and B′ B′ daughter nuclei. One of the A′ nuclei passes through this specially formed clamp connection so that it lies between the two B′ nuclei. The four nuclei then ie along the length of the narrow cell in the pattern A′ B′ A′ B′ and a cell wall is formed dividing them into two cells,

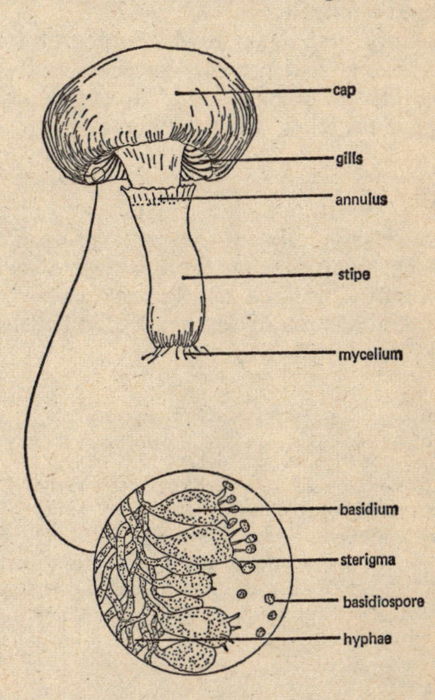

Fig. 15.—The detailed structure of *Agaricus campestris*.

each with A′ and B′ nuclei. The secondary mycelium formed from these cells produces a mushroom or *basidiocarp* with folds of tissue on its underside called *gills* on which *basidia* develop. These are binucleate club-shaped cells at the ends of the hyphae. The two nuclei fuse forming a diploid nucleus and meiosis then takes place to produce four haploid nuclei. Each of these nuclei passes through a short tube into a little sac to become a *basidiospore*, so that the four basidiospores are like knobs at the end of the four spikes protruding from the end of the basidium (*see* Fig. 15).

NOTE: In the cultivated mushroom, only two basidiospores develop.

The basidiospores are shot off from the basidium with the aid of a drop of water and can germinate on suitable material to produce a primary mycelium.

Other basidiomycetes have much more complex life cycles which involve the asexual production of many different kinds of spores such as uredospores and teleutospores. Basidiomycetes which cause rust diseases of many crops form such spores.

30. Economic importance. Mushrooms are grown commercially for food although many other mushroom-like fungi are edible. Others are not and some produce very virulent poisons. Both rust and smut diseases of crops are caused by basidiomycetes and include some of the worst plant diseases known, especially on cereal crops.

PROGRESS TEST 2

1. How are plants classified? (1)
2. What shape, size and structure are viruses? (3–4)
3. How do viruses reproduce? (5)
4. What is the economic importance of viruses? (6)
5. What are endospores and what is their function? (8)
6. What are the types of nutrition found in bacteria? (9)
7. How do bacteria reproduce? (13)
8. How are fungi classified? (20)
9. What is the typical life cycle of an ascomycete? (24)
10. Describe the life cycle of the common mushroom *Agaricus campestris*. (29)
11. What is the economic importance of the basidiomycetes? (30)

CHAPTER III

THE ALGAE

1. General features of the algae. The algae include the green limes on ponds, green films on trees and walls, the seaweeds f the shores and the plant plankton that support animal life ı the seas. Algae were responsible for producing the oil eserves of the world.

They are relatively simple plants which have cellulose cell 'alls and contain chlorophyll, although the green colour may e masked by other pigments. They are photosynthetic utotrophs. Most are aquatic and even the terrestrial forms equire a film of water for the movement of the reproductive ells.

2. Reproduction in the algae. The male and female *gametes* r reproductive cells may be of a similar size and free-swimming, ı which case the algae are said to be *isogamous*. Others have ıore advanced reproductive processes and the free-swimming ametes are of different sizes. Such algae are *heterogamous*. 'he most advanced algae have a non-motile female gamete vhich is retained within the body or thallus of the alga while he male gamete is free-swimming. These algae are *oogamous*. ıs well as showing a trend towards more advanced forms of eproduction, the algae also demonstrate a wide range in size nd development, from single-celled bodies through undifferentiated multicellular bodies to differentiated multicellular odies, although these do not have a developed vascular ystem.

3. Classification. Algae are grouped with the bacteria and he fungi into the division of the plant kingdom called the Challophyta. Within the algae there are eleven different lasses, but only seven are of importance. These are:

(*a*) *Chlorophyceae* or green algae;
(*b*) *Cyanophyceae* or blue-green algae;

(*c*) *Bacillariophyceae* or diatoms;
(*d*) *Dinophyceae* or dinoflagellates;
(*e*) *Euglenophyceae* or euglenoids;
(*f*) *Phaeophyceae* or brown algae;
(*g*) *Rhodophyceae* or red algae.

CHLOROPHYCEAE (*GREEN ALGAE*)

4. The green algae are bright green in colour and show a wide range of forms from the unicellular *Chlamydomonas* and the multicellular filaments such as *Spirogyra* to multicellular sheets up to 30 cm in length such as the sea lettuce *Ulva lactuca*. Most are aquatic but the ability of a few to live out of water suggests that they may be related to the ancestors of the higher plants. Some green algae are constituents of lichens (*see* **27**). Those green algae found on sea shores tend to be found quite high up the shores, towards the high tide mark

5. Chlamydomonas is an example of a unicellular freshwater alga. Figure 16 shows its structure to be microscopic in size with two anterior *flagellae* which propel it along with a screw

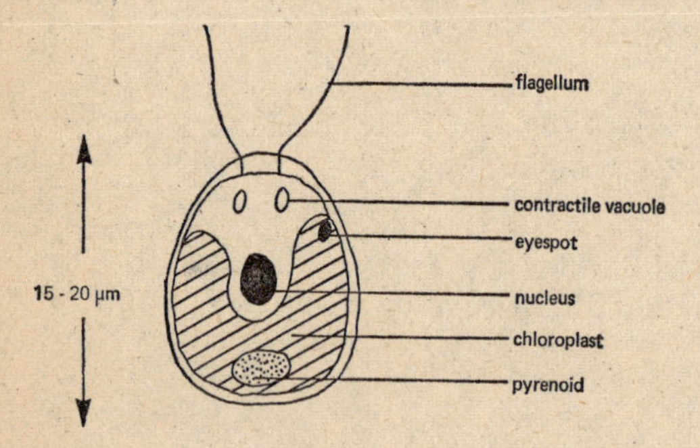

FIG. 16.—The detailed structure of a green alga—*Chlamydomonas*.

like motion. Two *contractile vacuoles* control the water and salt content of the cell. These fill up and discharge their contents to the exterior. The *eyespot* is a light-sensitive area of the cell able to direct the movement of the cell towards light. The chloroplast is bowl-shaped and has a *pyrenoid* set in it which is a proteinaceous body which stores the starch produced by photosynthesis, *Chlamydomonas* being an autotrophic organism. The nucleus contains one set of chromosomes and is therefore haploid.

6. Reproduction can be asexual or sexual.

(*a*) *Asexual reproduction.* Under favourable conditions, mitosis takes place and four *zoospores* are produced which are released and develop into full sized cells.

(*b*) *Sexual reproduction.* Under unfavourable conditions mitosis continues and results in sixty-four gametes which are released into the water. The gametes are thus isogamous and two originating from different parents join up and the nuclei fuse to give a diploid *zygote.* This develops a resistant cell wall which helps it to withstand adverse conditions. It ultimately undergoes meiosis and when the wall eventually breaks down four haploid zoospores are released which grow into four new cells.

CYANOPHYCEAE (*BLUE-GREEN ALGAE*)

7. The blue-green algae are mainly found in water where there is a lot of organic matter. Some of them are used as indicators of water pollution, others are able to survive high temperatures and inhabit hot springs, and others are constituents of some lichens. *Nostoc* is a blue-green alga which is able to fix nitrogen. It is common in water in rice paddy fields where it helps to maintain fertility from year to year.

8. Size, shape and structure. Blue-green algae are all either microscopic in size or just visible to the naked eye. They range from single cells to filamentous colonies. The cells have no true nucleus, endoplasmic reticulum or mitochondria but they do have ribosomes and chlorophyll and obtain their food by photosynthesis, even though the green colour of the chlorophyll is often masked by the blue-green pigment *phycocyanin.*

9. Reproduction. Asexual reproduction occurs through fragmentation of the filaments or by division in the unicellular species. Mitosis does not occur as the nucleus is not surrounded by a nuclear membrane and is ill-defined. Sexual reproduction has not so far been observed in the blue-green algae.

BACILLARIOPHYCEAE (DIATOMS)

10. Diatoms are found in both salt and fresh water and can be unicellular or colonial. Even colonial diatoms are microscopic in size. The cell wall, which is composed of silica and pectin, is made up of two parts or *valves* which are rather like a shallow box with a deep overlapping lid, as shown in Figure 17.

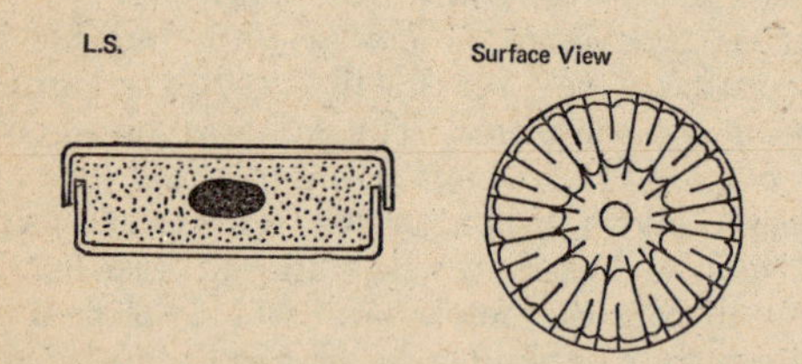

FIG. 17.—The arrangement of valves in a diatom. L.S. is a longitudinal section.

The surface of the valves is often highly sculptured and ornamented with pores through which the protoplast is in contact with the water. The cells may be either radially or bilaterally symmetrical. Diatoms have a well-defined nucleus but the chlorophyll is masked by yellow-brown pigments.

11. Reproduction. Both asexual and sexual reproduction have been observed in diatoms.

(*a*) *Asexual reproduction* is brought about by the division of the nucleus and the cytoplasm and the separation of the two valves. Each new daughter cell then secretes a new valve which is always the equivalent of the "bottom of the box". After a few generations, the diatoms have diminished markedly in size and at a certain point both valves are shed and the protoplast grows to full size again and secretes two new valves.

(*b*) *Sexual reproduction* is by *conjugation*, the union of two gametes to form a zygote which undergoes meiosis prior to germination.

12. Economic importance. Diatoms form a large part of the *phytoplankton* or floating plants on which aquatic animals

argely depend for food. The silica in the walls of diatoms does
ot decay when diatoms die, and where large accumulations of
'skeletons" occur in some parts of the world, diatomaceous
arths are exploited, the material being of value as an abrasive,
or insulation and as a constituent of filters.

DINOPHYCEAE (DINOFLAGELLATES)

13. **Dinoflagellates** are found in both marine and fresh
vater environments. They are all unicellular and motile,
notility being achieved by the use of two flagellae which are
et in grooves around the cell. Often the cells have no walls
ut when present they may consist of cellulose plates. Dino-
lagellates contain a yellow-brown pigment as well as chloro-
hyll. Reproduction is mainly by asexual cell division.
)inoflagellates are important constituents, along with diatoms,
f phytoplankton. Many of them are luminescent.

EUGLENOPHYCEAE (EUGLENOIDS)

14. **The euglenoids** are mainly found in fresh or brackish
vater. They are especially common in nitrogen-rich water
uch as farm-yard ponds.

15. **Size, shape and structure.** They are between 50 and 150
m in length and are spindle-shaped with one anterior flagellum
roviding motility. There is no cell wall but the cells are sur-
ounded by a *pellicle* which covers fibres or *myonemes*. These
ibres can, by their differential contraction, change the shape
f the organism. The cell contains a nucleus and chloroplasts,
he latter varying greatly in shape in different species. A
ontractile vacuole regulates the water content of the cell and
nay aid removal of waste materials. There is a light receptor
ear the base of the flagellum and an *eyespot* set in the cyto-
lasm near the light receptor which it is thought to shade.

16. **Nutrition and reproduction.** In green euglenoids food is
nade by photosynthesis, but some euglenoids are colourless
nd these are able to either absorb soluble food or even engulf
olid particles. Reproduction is by division of the cell along its

length to form two daughter cells. Sexual reproduction has not been recorded.

NOTE: Euglenoids have been intensively studied for their metabolic processes. They show many animal qualities and are often classified as animals as well as plants.

PHAEOPHYCEAE (BROWN ALGAE)

17. The brown algae are almost entirely marine organisms and range from dark brown to olive green in colour, the chlorophyll being masked by the brown pigment fucoxanthin. The common brown seaweeds, kelps and wracks found in intertidal zones on rocky shores range in size from a few centimetres to 30 metres for some large kelps. The brown coastal seaweeds tend to be intermediate in location between the green algae high up the shore and the red algae lower down. The plant body of the brown algae is diploid with two sets of chromosomes in each cell nucleus.

18. Fucus vesiculosus, or bladder wrack, is one of the most common seaweeds on rocky shores in Britain. Its size varies according to the exposure of the habitat but average size is about 40 to 50 cm.

(*a*) *Shape*. This is given in Fig. 18, which shows the bladder wrack to be made up of flattened branches which are attached to the rock by a rounded stem called a *stipe*, the end of which is flattened for better adhesion and is called a *hapteron* or *holdfast*. Along the flattened branches are oval air bladders which give the branches or fronds buoyancy. The ends of many of the branches are swollen and contain the flask-shaped *conceptacles* in which the gametes are produced. These open on to the surface of the frond through a small pore, the *ostiole*. The swollen ends of the branches are termed *receptacles*.

(*b*) *Structure*. A section across the branch of the thallus shows it to be made up of three layers:

(*i*) *An outer layer* of box-like cells which provide protection and absorb light for photosynthesis;

(*ii*) *A cortex* of rounded cells for storing *fucosan*, the product of photosynthesis in *Fucus*;

(*iii*) *A medulla* of elongated cells, some of which show

rudimentary sieve elements like those found in more highly evolved plants and which are concerned with transport within the plant.

(*c*) *Nutrition.* *Fucus*, like other brown algae, produces its own food by photosynthesis.

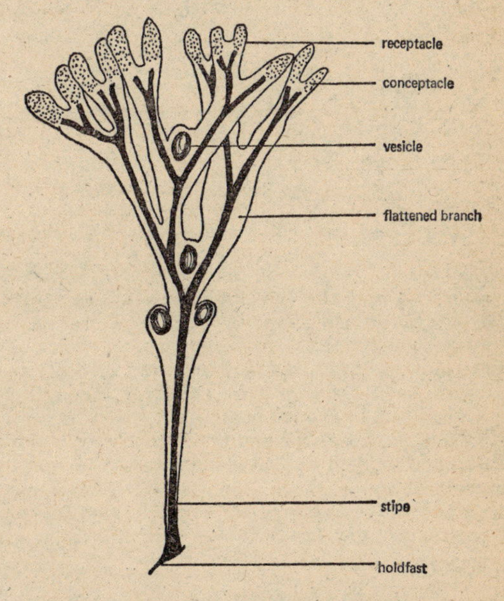

FIG. 18.—*Fucus vesiculosus*—the bladder wrack.

19. Reproduction.

(*a*) *Asexual reproduction.* This is rare, but occasionally outgrowths are produced which develop into a new plant. However this is not considered a reliable method.

(*b*) *Sexual reproduction.* *Fucus* seaweed is diploid with two sets of chromosomes in all the cells except the gametes, which are haploid. The conceptacles of one plant produce only male or female gametes and *Fucus vesiculosus* is therefore termed *dioecious*. Other *Fucus spp.* are *monoecious* in that they produce male and female gametes in conceptacles on the same plant.

(*i*) *The female conceptacle* (*see* Fig. 19) contains sterile hairs called *paraphyses* and developing oogonia. The oogonia develop from a single cell which pushes up from the floor of

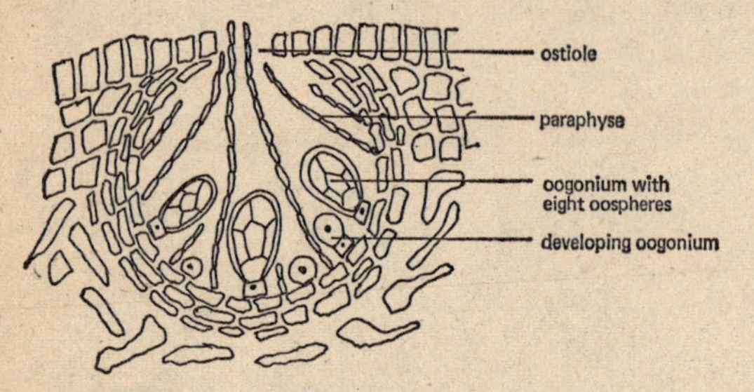

Fig. 19.—The female conceptacle in *Fucus vesiculosus.*

the conceptacle. A division takes place to produce two cells, the stalk cell and the developing oogonium and a meiotic and a mitotic division then take place to produce eight haploid nuclei which are protected by two cell walls. The nuclei develop into eight oospheres. The outer cell wall releases this bundle of oospheres which are extruded through the ostiole with the aid of *mucilage*, a gluey liquid. Once outside the conceptacle the oosphere bundle is dispersed in the sea. The last wall is dissolved by the seawater and the oospheres sink to the sea bed.

(*ii*) *The male conceptacle* (*see* Fig. 20), also has paraphyses but the antheridia are produced on branched structures.

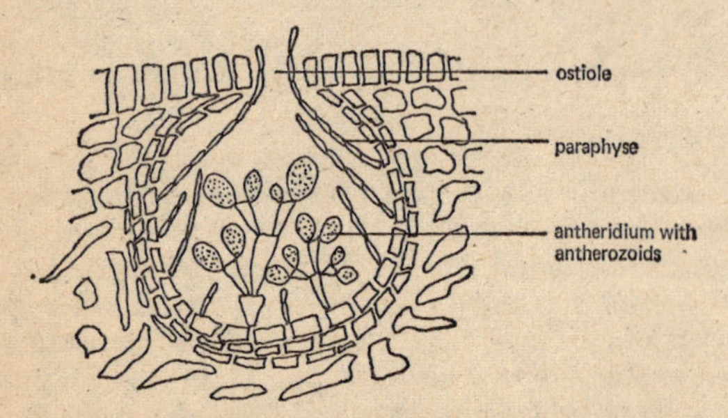

Fig. 20.—The male conceptacle in *Fucus vesiculosus.*

Each antheridium is a sac-shaped branch which contains sixty-four haploid antherozoids produced by meiosis and mitosis. The antherozoids are kidney-shaped with two flagellae which make them motile and have an eyespot which is light sensitive. They swim out of the conceptacle, at which point the eyespot detects light and the antherozoids swim away from it to the bottom of the sea, where they come into contact with oospheres.

One antherozoid then fertilises an oosphere to give an *oospore*. This divides to produce a new plant or thallus which may be either male or female.

20. Economic importance. The coastal brown algae are used as fertilisers by farmers as they provide the salts of nitrogen and potassium as well as humus. In Asia they are used as a food and have also been used in many countries as a source of iodine. Brown algae are a major source of *alginates*, colloidal substances used to thicken ice-cream and cosmetics and also used in pharmaceuticals.

RHODOPHYCEAE (*RED ALGAE*)

21. The red algae are mainly marine although a few species are adapted to fresh water. They are found in the intertidal zone, usually well down the shore but they can also live deep below the surface of the sea, down to 45–55 metres around the coasts of Britain and down to 180 metres in the tropics. They are therefore well adapted to photosynthesise in very poor light conditions.

22. Size, shape and structure. They are not normally more than 1 metre in length. The thallus may be cord-like, flattened or branched and feathery, and some unicellular red algae are known. They range in colour from pink to dark red and purple, the colour being due mainly to *phycoerythrin*. Many red algae accumulate lime and are therefore hard and crusty.

23. Reproduction. The thalli of the red algae are haploid and may be monoecious or dioecious. There are no motile cells in the red algae; even the male gamete *spermatium* is non-motile, relying on sea currents for transport. The haploid spermatium, produced by mitosis, fuses with the nucleus of the haploid female structure, a flask-shaped *carpogonium*, to give a diploid *carposporophyte*. This is retained on the haploid plant and produces *carpospores* which give rise to another diploid generation, the *tetrasporophyte*, which is separated from the haploid thallus and lives freely. It then undergoes meiosis and the haploid tetraspores so produced give rise to the new haploid thalli.

24. Economic importance. The ability of red algae to accumulate lime makes them important reef producers in warm seas. They are collected, like the brown algae, for their alginates and produce a jelly-like substance called *agar* which is very widely used for making media for growing micro-organisms. Red algae, recognised locally by names like Irish Moss, Carrageen, Laver and Dulse, are eaten as foods in parts of Northern Europe.

LICHENS

25. The lichens. A lichen consists of an alga and a fungus living together, and lichens form a group separate from both algae and fungi. They are found growing on the bark of trees, on decaying wood, on rocks and walls and on the surface of soil.

26. Size and shape. Lichens may be microscopic or very small but they can grow to 30–60 cm in diameter. They only grow a few millimetres each year. There are four main types of lichen:

(*a*) *crusty, often bright yellow lichens* found on rocks and walls;
(*b*) *greenish-grey upright lichens*, many of the upright parts widening towards the end into small cups;
(*c*) *shaggy foliose growths* usually hanging from branches of trees;
(*d*) *quite large leafy growths* found on walls and tree trunks.

27. Structure. A lichen is made up of two components:

(*a*) *a fungus* which is usually an ascomycete but occasionally a basidiomycete;
(*b*) *an alga* which is either a green or a blue-green alga.

The fungus cannot live by itself but the alga can live independently of the fungus. The fungus forms the framework of the thallus and the rhizoids which fasten the lichen to the substratum of rock or soil. The algal cells are set in pockets in the fungus and are penetrated by branches of the fungus called haustoria.

28. Nutrition. The alga, with its ability to photosynthesise, provides the fungus with organic food and in return receives water, essential mineral salts and shelter. Lichens are examples of *symbiosis*, where two organisms live very close together for the benefit of them both.

29. Reproduction.

(*a*) *Asexual reproduction* is more common than the sexual form in lichens; it may take place in one of two ways:

(*i*) *a piece breaks off the thallus* and develops into a new plant;
(*ii*) *soredia* are produced which are balls made up of algal cells intertwined and penetrated by the fungus. They are produced inside the lichen and are then pushed to the surface and released. After dispersal they develop into new lichens.

(*b*) *Sexual reproduction.* Here the fungus produces spores which are dispersed. They germinate and produce filaments which trap free-living algal cells and the two develop into a new lichen. If the fungal filaments fail to come into contact with the alga they die, so this method is rather an unsatisfactory procedure and probably of minor importance in lichen reproduction.

30. Economic importance. Lichens have several economic uses:

(*a*) *they provide important grazing material* for herbivores such as reindeer and moose in the tundra of the far North;
(*b*) *lichens provide natural dyes* which produce more muted colours than the harsh colours of many synthetic dyes;
(*c*) *lichens are the source of litmus*, the pigment which is used in chemistry because of its ability to turn red under acid conditions and blue under alkaline conditions;
(*d*) *lichens are used in the perfume industry.*

31. Ecological significance. Lichens are important in two main ways.

(*a*) *They are very hardy and resistant to drought* and man species will colonise bare rock. As such, they commence th colonisation of such inhospitable environments and eventu ally allow higher plants to grow as well (*see* IX, **10**).

(*b*) *Lichens are very susceptible to sulphur dioxide* pollutio in the air (*see* XII, **3**). Because of this they can be used a *biological indicators* of such pollution and the presence c absence of lichens in and around urban and industrial area can be useful in mapping variations in sulphur dioxid pollution.

PROGRESS TEST 3

1. What are the general features of the algae? **(1)**
2. How do algae reproduce? **(2)**
3. How are algae classified? **(3)**
4. What is the structure of *Chlamydomonas*? **(5)**
5. What are the main features of the blue-green algae? **(8)**
6. What is the economic importance of the diatoms? **(12)**
7. Why is *Euglena* classified both as an animal and a plant **(15–16)**
8. How does *Fucus vesiculosus* (bladder wrack) reproduce **(19)**
9. How does reproduction in the red algae differ from that i other algae? **(23)**
10. What are lichens? **(25–27)**
11. How do lichens reproduce? **(29)**
12. What is the ecological significance of lichens? **(31)**

CHAPTER IV

LIVERWORTS, MOSSES AND FERNS

BRYOPHYTA (*LIVERWORTS AND MOSSES*)

1. Introduction to the bryophytes. The bryophytes live on land but most of them prefer moist shady places and they require water for the movement of the gametes. They are fairly simple plants with little internal differentiation except in the reproductive structures. The bryophytes normally take in water through the whole body and through hairlike processes called *rhizoids*.

2. Reproduction in the bryophytes. There are two distinct phases in reproduction. The plant body is haploid and is the *gametophyte*. It produces the gametes by mitosis and the female gamete is retained in the plant for protection. The male gamete fuses with the female gamete to give a diploid *sporophyte*. This sporophyte generation, which is attached to the gametophyte, produces haploid spores by meiosis and these spores develop to give a new gametophyte generation. This is termed the *alternation of generations* and is represented diagrammatically in Fig. 21.

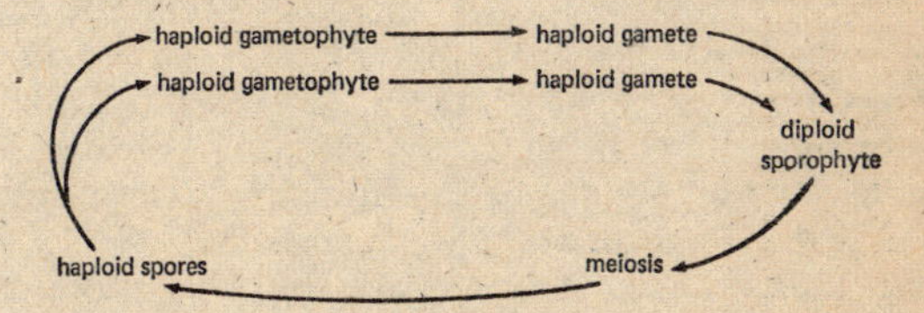

Fig. 21.—A generalised bryophyte life cycle, showing the alternation of generations.

3. Classification. There are two classes in the bryophytes:

(*a*) *Hepaticae* or liverworts;
(*b*) *Musci* or mosses.

The liverworts have *elaters*, sterile cells produced in the sporophyte along with the spores. These do not occur in the mosses but the mosses have a much better developed *water conduction* system than the liverworts.

HEPATICAE (LIVERWORTS)

4. **Pellia** is one of the simpler liverworts and is most commonly found in damp places especially underneath the overhanging banks of streams and ditches.

5. **Size, shape and structure.** Each *Pellia* plant is about 10 to 15 cm in length and consists of flattened branches which are attached to the ground by unicellular rhizoids. It is dichotomously branched so that each branch divides into two equal branches. These branches are interwoven with branches from other plants.

The plant is surrounded by a *cuticle* to stop desiccation and beneath the cuticle are rounded cells with chloroplasts, these cells so being arranged that the centre of the branch is thicker than the edges.

6. **Nutrition.** Photosynthesis takes place in the cells containing chloroplasts, *Pellia* being a photosynthetic autotroph even though it inhabits very shady places. Water is taken in through the whole plant and through the rhizoids, as are mineral salts. The main function of the rhizoids is anchorage rather than absorption.

7. **Reproduction.** Asexual reproduction is not found in *Pellia* although it does occur in other liverworts. Balls of cells called *gemmae* are produced in *gemmae cups* which are depressions on the upper surface of the branches. The gemmae are dispersed and then grow to give new plants.

Sexual reproduction takes place by the production of haploid male gametes in *antheridia* (*sing.* antheridium) and haploid female gametes in *archegonia* (*sing.* archegonium) by the *gametophyte plant* (*see* Fig. 22).

(*a*) *The gametophyte antheridium.* About half way along the branches of the plant near the central thickened area are found cup-shaped depressions with stalked antheridia inside.

Within these antheridia are found the *antherozoid mother cells* which divide by mitosis to give motile *antherozoids*. The long tapered body of the antherozoid is coiled and attached to the narrower end are two long flagellae which provide motility. When they are mature, the antherozoids burst out of the antheridia on to the surface of the plant.

(*b*) *The gametophyte archegonium.* These are produced in groups near the end of the branch and are protected by a covering flap of tissue known as an *involucre*. Only one of each group of archegonia develops. Each archegonium is flask-shaped and the lower, wider part which is attached to the plant is called the *venter*. Within the venter are found the *oosphere* and the *ventral canal cell*. The upper narrower part is the tubular *neck* and the top is covered with four cells forming a *cap*.

When the archegonium is mature, the cells lining the neck and the ventral canal cell disintegrate and form mucilage. This pushes off the cap cells and the sugar it contains attracts the antherozoids, a process known as *chemotropism*. The antherozoids swim in a film of water to the archegonium and one fertilises the oosphere to give a diploid *oospore*.

8. The sporophyte. The oospore divides mitotically to produce the diploid *sporogonium* which grows and results in a basal part still attached to the *Pellia* plant called a *foot*, a central stalk called a *seta* and a terminal part called a capsule or *theca* which is covered by a *calyptra*.

Within the theca are found *spore mother cells* and sterile cells called elaters. The spore mother cells divide by meiosis and produce haploid spores but the sporogonium remains dormant for a few months. In spring, the seta elongates and the theca ruptures the calyptra and is carried up on a stalk 5 to 8 cm in length. The capsule bursts open into four parts and the spores are liberated through the action of the elaters which twist and stir up the spores. During the dormant period the spores have divided and when liberated they are multicellular and contain chlorophyll. They start to grow immediately to form a new gametophyte plant.

9. Economic importance. Although widespread, there are no major economic uses for liverworts. They grow in places such as very damp shady stream banks which would otherwise be unoccupied by other plants.

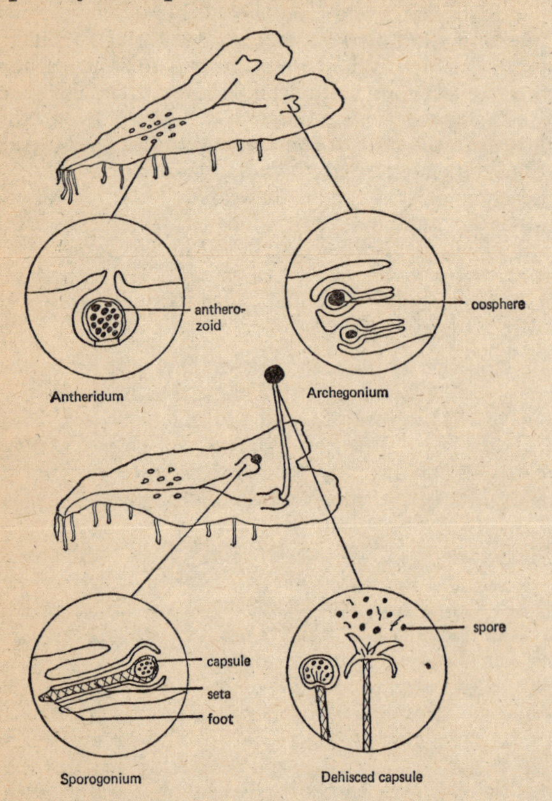

FIG. 22.—A liverwort—*Pellia*—showing the sporophyte and the gametophyte plant from which it develops.

MUSCI (*MOSSES*)

10. The mosses are normally erect plants and are radially symmetrical having leaf-like structures attached around a central stem. They are fairly successful land plants but still require water for reproduction.

11. Mnium hornum is a common moss found in large tufts in beech woods and on shady banks.

12. Size, shape and structure. The haploid gametophyte plant is 5 to 8 cm in height and has a brownish-green stem with spirally arranged bright green leaves. Branched rhizoids attach the plant to the ground and may give rise to further plants. The internal structure of the stem shows some differentiation:

(*a*) *the axial cylinder* is in the centre of the stem and is made up of elongated thin-walled *hydroid* cells which are responsible for water conduction although a lot of water is carried up the plant by capillary action on the surface;
(*b*) *the cortex* is composed of larger cells containing chloroplasts for photosynthesis together with starch as a stored product.
(*c*) *the epidermis* around the outside consists of two layers of thick-walled cells which provide protection.

The leaves are made up of a *midrib* which contains conducting tissue which in some mosses is continuous with that of the stem, and a *lamina* which is the thinner part of the leaf and consists of cells packed with chlorophyll for photosynthesis. Growth takes place by division of an *apical cell*.

13. Nutrition. *Mnium hornum*, like other mosses, is a photosynthetic autotroph. It collects mineral salts from the ground on which it is growing.

14. Reproduction.

(*a*) *Asexual reproduction* takes place in damp conditions when the rhizoids produce branched outgrowths upon which new plants develop.
(*b*) *Sexual reproduction*. In *Mnium* the male antheridium and the female archegonium are borne on separate plants, so it is *dioecious*. There are also monoecious mosses.

(*i*) *Antheridia*. The end of the stem is flattened and has a ring of leaves around the edge forming an *involucre*. Within the involucre are found twenty to fifty antheridia intermingled with sterile hairs, the *paraphyses*. The antheridia are oval structures on a short stalk. Antherozoids are formed in the antheridia by mitosis and are spiral cells with two flagellae. When mature they are liberated through the top of the antheridium and swim to the archegonia.

(*ii*) *Archegonia* are produced at the ends of shorter, darker female plants. They are protected by an inward-curling involucre. They are flask-shaped with a large basal ventre containing the female gamete, the oosphere. Above the ventre is the tubular neck, with ventral and neck canal cells, which is closed by a cap until the oosphere is mature. Then some sucrose-containing mucilage is produced which pushes off the cap and attracts antherozoids. Paraphyses are found dispersed among the archegonia.

Fertilisation takes place in May in Britain when the plants are covered by a film of water in which the antherozoids swim. The paraphyses are able to secrete water and contribute to the production of the water film. After fertilisation the oosphere forms a diploid oospore and this develops to produce a sporogonium which is the sporophyte generation.

15. The sporophyte generation. Usually only one sporangium develops on each plant and this takes about 11 months. When complete, it is composed of three parts:

(*a*) *The foot*, set deep in the gametophyte tissue, which absorbs water and mineral salts for the sporogonium;

(*b*) *The seta* or stalk which is short at first but grows later to about 3 cm;

(*c*) *The capsule* or *theca* in which the spores are produced.

The capsule is at first covered by a *calyptra* which is the remains of the archegonium, but this is shed to show the elongated capsule. Its lower end is called the *apophysis* and has *stomata* and many chloroplasts and provides much of the food for the sporogonium.

The middle of the capsule has a core of sterile cells forming a *columella* around which are the cells of the archesporium. These archesporium cells undergo meiosis and produce the dark green haploid spores which have a store of oil. Around the archesporium is loose filamentous tissue which disintegrates when the spores are ripe. The outer wall of the capsule is made up of several layers of chloroplast-containing cells and an *epidermis*.

The cone-shaped top of the capsule is found to be covered by a circular lid or *operculum* when the calyptra falls off. The edge of the operculum is attached to the capsule by a ring of specialised cells called the *annulus*. When the spores are ripe, the cells of the annulus rupture and the operculum is shed to

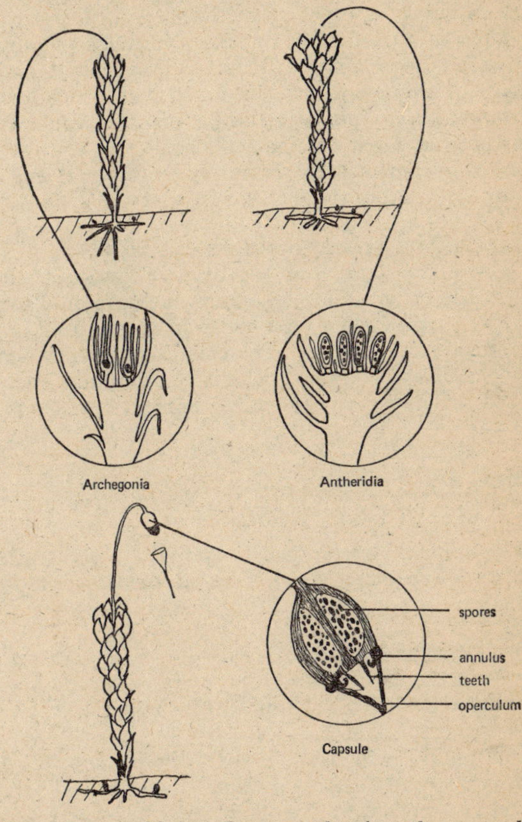

FIG. 23.—A moss—*Mnium hornum*—showing the sporophyte and gametophyte generations.

expose the *peristome*. This is a double ring of hygroscopic teeth which are able to move in response to changes in humidity.

The ripe spores become loose in the capsule which now bends over so that the peristome is facing downwards. When the air is dry, the peristome teeth curl back and spores are flung out, but as the air becomes moist, they curl back inwards again, closing the capsule. Spores are therefore only liberated in dry conditions.

After dispersal the spores remain dormant until wet weather prevails when they germinate to form a branched filamentous structure called the *protonema*. This develops rhizoids and buds at the base of the branches, the buds developing into new haploid gametophyte plants. All the plants formed from a particular protonema are of the same sex.

Mnium, like other mosses, thus fully exhibits alternation of generations, rather like the liverworts.

16. Economic importance. Mosses can be weeds of damp, badly drained grassland and are of very limited value for grazing. On the other hand, sphagnum moss, mainly found in wet conditions, can form a peat which is widely used in horticulture. Mosses are also useful as colonisers of rocky ground after lichens. They are initial colonisers of ground made bare by fire, and are useful in preventing erosion of soil following fires.

PTERIDOPHYTA (FERNS, HORSETAILS AND CLUB MOSSES)

17. The pteridophytes are generally well adapted to live on the land, although many species are adapted to damp shady conditions.

18. Size, shape and structure. They range in size from a few centimetres in the club mosses to 1–2 metres in the horsetails found in Britain and to 7–20 metres in height in the case of large tree-like tropical ferns.

All these plants, which are clearly divided into stem, leaf and root, have a well developed vascular system. They are the sporophyte generation, unlike the bryophytes where the main plants are the gametophytes. In the pteridophytes, the gametophyte generation is separated from the sporophyte and is small and insignificant. The pteridophytes are divided into:

(*a*) *the Filicales* or ferns;
(*b*) *the Equisetales* or horsetails;
(*c*) *the Lycopodiales* or club mosses.

FILICALES (FERNS)

19. The ferns, with the exception of tree ferns, have an underground stem or *rhizome* with only the top appearing at ground level. The crowded leaves, called *fronds*, are spirally arranged at the tip of the rhizome. The spores are borne on the undersides of the fronds and after dispersal they germinate to produce the *prothallus* (gametophyte generation). These are similar in all the *Filicales*. A typical fern is the male fern, *Dryopteris filix-mas*.

20. Dryopteris filix-mas is widely distributed in Britain and is most commonly found in woods and hedgerows.

21. Size, shape and structure. This fern may be up to a metre or more in height. It has an underground stem or rhizome which is well supplied with roots and from the tip of which, at ground level, arise the leaves or fronds. These are rolled up when young but unroll to disclose a stiffened rod-like *rachis* to which are attached leafy structures called *pinnae*. The pinnae are large at the bottom and decrease in size to the tip. They are subdivided further into *pinnules*.

NOTE: The growing end of the rhizome and the young curled fronds are covered with soft brown scales called *ramenta*.

The conducting or vascular tissues are very similar to those of flowering plants (*see* VI, **9-10**) except that the *xylem* consists mainly of *tracheids* without any vessels. The vascular tissues of the root are continuous with those of the rhizome. In the rhizome they form a tubular network or *dictyostele* to which the vascular tissues of the fronds are attached.

There is no *cambial tissue* in *Dryopteris* and so there can be no secondary thickening. Extra support is supplied by a ring of *sclerenchyma* cells.

22. Nutrition. The pinnae are well supplied with chloroplasts and stomata to facilitate photosynthesis. Water and mineral salts are taken up from the soil by the roots.

23. Reproduction.

(*a*) *The sporophyte generation.* The fern plant is the diploid sporophyte generation and produces haploid spores by meiosis. The spores are borne in sporangia which are found in patches or *sori* (*sing.* sorus) on the underside of the

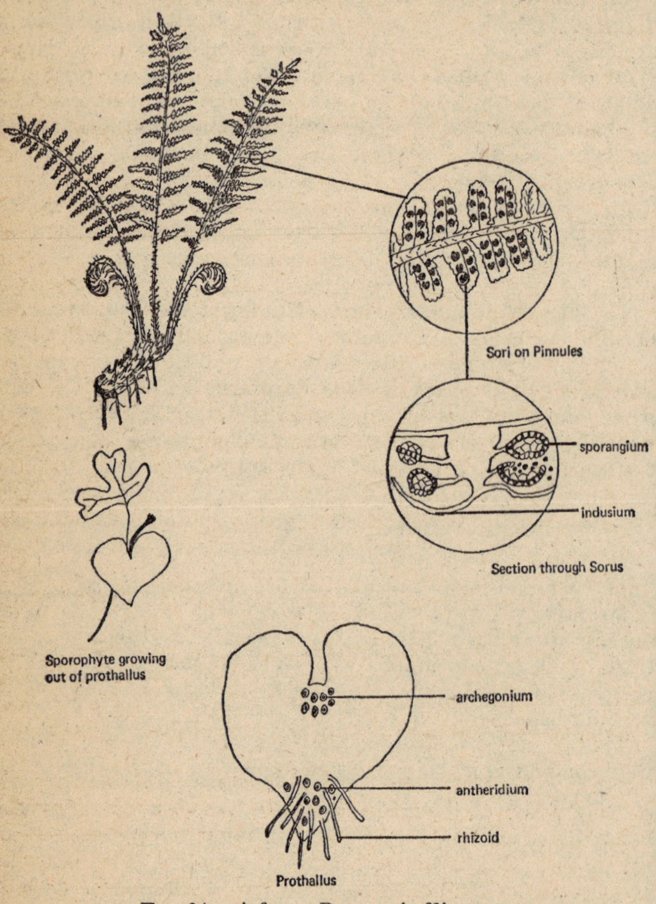

Fig. 24.—A fern—*Dryopteris filix-mas*.

pinnae. The sori have a membraneous covering called an *indusium*. The sporangia have a short stalk and a disc-shaped head which has thin-walled sides, which are in turn joined together by a strip of thickened cells called the *annulus*. At one side the thickened cells are replaced by thin walled cells forming a pore or *stomium*. The sixty-four haploid spores are produced from spore mother cells. When they are ripe the indusium dries and curls back to expose the sporangia. Further drying of the sporangia sets up tension in the annulus which causes the stomium to tear open and eject the spores. These germinate to produce the gametophyte generation.

(*b*) *The gametophyte generation*. The spores germinate to produce an independent heart-shaped *prothallus* which is less than 1 cm across. The prothallus has a cushion-like thick centre and thin edges, is green in colour and has rhizoids which attach it to the ground. Its underside bears both antheridia and archegonia, the former found near the tip and the latter set in the cushion near the notch of the heart-shaped prothallus.

The antheridia are suspended below the prothalli and have thin walls. They contain many coiled flagellate antherozoids, the end of the antheridium being pushed off to release them once they are mature. The archegonia are flask-shaped with the ventre set deep in the prothallial tissue and the curved neck protruding. The ventre contains the oosphere whilst the neck contains cells which later disintegrate to form mucilage. This serves to attract the antherozoids when the oosphere is mature. One antherozoid fertilises the oosphere which then becomes a diploid oospore.

The oospore divides and grows into the sporophyte plant. It remains attached to the prothallus and is at first dependent on it for its food but later the sporophyte becomes independent and the prothallus dies and disintegrates.

24. Economic importance. Some ferns are important weeds, especially of upland pasture, as they have virtually no grazing value and are difficult to destroy, partly because of the persistent nature of the underground rhizome. A few ferns are propagated for their ornamental value. Ferns were important in the formation of coal deposits.

EQUISETALES (HORSETAILS)

25. The horsetails. Most species of horsetails today are limited to waste, gravelly ground, but ancestors of present-day plants were important constituents of the forests of Carboniferous times over 300 000 000 years ago and included large tree-like plants.

26. Equisetum arvense, the field horsetail in Britain, is a typical member of the *Equisetales*. It commonly occurs as a weed in fields and on waste ground and in hedgerows.

27. Size, shape and structure. It is about 20 to 80 cm high and has a stiff upright jointed stem to which branches are attached in whorls. Leaves are very small and are fused to form a sheath around the stems. These stems are green and are the main photosynthetic area of the plant. There are, however, other brown stems which have cones called *strobili* at the ends. These are the reproductive structures. The upright stems are attached to an underground stem or rhizome from which the roots arise. The rhizomes may develop swollen tuberous ends which can act as organs of vegetative propagation.

28. Reproduction.

(*a*) *The sporophyte generation* is the horsetail plant and is diploid. It produces haploid spores by meiosis in sporangia in the strobili. The strobilus is composed of whorls of mushroom-like outgrowths called *peltate scales*, to the underside of which are attached the sporangia. When ripe the spores are liberated from the sporangia. They have curled strips of tissue attached to them called elaters which are hygroscopic and help in dispersal. The spores are all more or less alike although they germinate to produce two different types of prothalli. One produces only male antheridia whilst the other is *hermaphrodite* and produces both antheridia and archegonia.

(*b*) *The gametophyte generation.* The male prothallus is small and irregular in shape with the antheridia produced in outgrowths on the upper side. The antherozoids are curled and have a number of flagellae for motility. The hermaphrodite prothallus has branches, on the ends of which are produced the antheridia, whilst later on the archegonia are produced at the bases of these branches. The archegonia are embedded in the upper side of the prothallial tissue and have long curved necks which open widely for the entry of the antheridia. The oosphere is deep within each archegonium, and after fertilisation becomes a diploid oospore and develops to produce the sporophyte plant.

29. Economic importance. They are rarely used by man, but can be harmful weeds and are very difficult to eradicate because of the underground stems and the tubers. Coal deposits were formed partly from the remains of prehistoric *Equisetales* (*see* **34** below).

LYCOPODIALES (CLUB MOSSES)

30. The club mosses have small leaves with membraneous outgrowths called *ligules*. The sporangia are grouped into cones or strobili. The spores may be all of one kind as in *Lycopodium* or of two types as in *Selaginella*. In temperate countries they are most common in peaty soil on mountains, but in the tropics other species grow on tree trunks.

31. Selaginella is a common greenhouse plant in Britain. It is a native of South America but has become naturalised in Cornwall.

32. Size, shape and structure. *Selaginella* is a creeping plant up to 30 cm long with many branches. On the branches are found pairs of alternately arranged leaves. The pairs are made up of a large and a small leaf, the smaller leaf always being on the upper side so that it causes less shading of the larger lower leaf. Each young leaf has a membraneous ligule at its base which withers as the leaf matures. At the base of each branch is produced a long root-like structure called a *rhizophore*. It is on the ends of the rhizophores that the true roots are found.

33. Reproduction. As with the other pteridophytes, the club moss plant is the diploid sporophyte generation. Most of the branches lie flat on the ground but the ones which bear the reproductive organs are vertical. The reproductive organs are called strobili. In a strobilus, the leaves are replaced by leaf-like structures called *sporophylls* which have sporangia attached to them. Each sporophyll has one sporangium attached to the upper side. Some are *megasporangia* which contain four large haploid *megaspores* and some are *microsporangia* which contain many haploid *microspores*. When the spores are ripe, the sporangia split open and the thick walled spores are flung a few centimetres away from the parent plant. The prothalli which are produced by the micro- and megaspores are very small, so much so that they never emerge from the spore cases. The microspores produce only antheridia whilst the megaspores produce archegonia.

The microspore contents divide into two cells before the spore is shed. Further divisions of one of these cells eventually produce the antherozoids which have elongated bodies and two flagellae. The megaspore also begins its development before it is shed. Divisions then take place inside the spore to produce female prothallial tissue and archegonia. The spore eventually splits open to expose the necks of the archegonia which now each contain an oosphere.

Fertilisation takes place by an antherozoid fusing with the oosphere to produce an oospore. The oospore divides into two cells, the embryo cell and the suspensor. The suspensor pushes the embryo cell into the centre of the megaspore prothallial tissue where it develops. Two leaves or *cotyledons* and a *radicle* are formed and later grow out to produce a new *Selaginella* plant. Thus an alternation of sporophyte and gametophyte generations is seen.

34. Economic importance. These plants are of very limited economic importance although a number of them are grown as ornamental plants. They were, however, very important in the formation of coal fields in Carboniferous times, over 300 000 000 years ago, and included giant plants such as *Lepidodendron*, nearly 1 metre in diameter and reaching to about 100 metres in height.

PROGRESS TEST 4

1. What are the main features of reproduction in the bryophytes? (2)

2. How does a liverwort obtain its food? (6)
3. What is the form of the gametophyte generation in liverworts? (7)
4. What is the size, shape and structure of the moss *Mnium hornum*? (12)
5. Describe the sporophyte generation in *Mnium hornum*. (15)
6. How are the *Pteridophyta* classified? (18)
7. How much tissue differentiation is there in ferns? (21)
8. Describe the gametophyte generation of a fern. (23)
9. What is the form of alternation of generations in the club mosses? (33)

CHAPTER V

THE GYMNOSPERMS

SPERMATOPHYTA

1. The spermatophyta or seed-bearing plants, which includ conifers and flowering plants show a much greater developmen of the vascular system than was found in the pteridophyte They also have their reproductive organs grouped into *cone* in the gymnosperms and *flowers* in the angiosperms or flowerin plants.

The *megaspore*, now called an *embryo sac*, is completel retained within the megasporangium and, along with th protective integument, forms an *ovule*. The fertilised ovul develops to produce a *seed*.

There are two subdivisions in the *Spermatophyta*.

(*a*) *Gymnosperms* where the seeds are naked, not enclose in an ovary (an additional protective tissue around th ovule).

(*b*) *Angiosperms* where the seeds are protected by a ovary which develops into a fruit.

GYMNOSPERMS

2. The gymnosperms are mostly fairly large plants, commonl trees or shrubs. The spore-bearing leaves or sporophylls ar grouped into cones, sometimes referred to as flowers. Th gymnosperms are *heterosporous*, producing two types of spore These are microspores, which develop into *pollen grains* an megaspores which develop into ovules. After fertilisation eac ovule matures into a seed.

There are a number of orders in the gymnosperms, many c which are only represented by fossils. Of the living gymnc sperms, the three most well-known groups are:

(*a*) *the Cycadales* or cycads;
(*b*) *the Ginkgoales* or Ginkgo tree;
(*c*) *the Coniferales* or conifers.

CYCADALES

3. Habitat. The cycads are mainly tropical or subtropical lants found in Africa, Australia and Eastern Asia.

4. Size, shape and structure. Cycads are palm-like plants ith a short trunk of soft loose wood and a head of radiating rge divided leaves. They may reach 10 metres but are more ommonly about 2 metres in height.

5. Reproduction. Cycads are dioecious with the cones being orne in the centre of the head of leaves. One Australian cycad reported to have female cones 1 metre in length and 20 cm ide. The pollen grains are carried by the wind but they erminate to form a *pollen tube* down which motile *sperm* or ntherozoids pass to fertilise the ovule.

GINKGOALES

6. Ginkgo biloba or the Maidenhair Tree is the only living epresentative of the Ginkgoes although fossil Ginkgoes have een recorded in the Permian era about 280 000 000 years ago. t was first recorded as a temple tree in China but is now grown n gardens all over the world.

7. Size, shape and structure. It is a pyramid-shaped tree which may reach 30 metres in height under good conditions. he flat leaves are fan-shaped with a single deep indentation. he tree, unlike most other gymnosperms, is deciduous, losing ts leaves at the end of a growing season.

8. Reproduction. Ginkgo is dioecious with male and female eproductive organs being borne on separate trees. The pollen rains are produced in a catkin-like structure composed of airs of microsporangia. The ovules are not borne in cones but ndividually or in pairs on long stalks.

After pollination by the wind, a germ-tube grows out of the ollen grain and motile antherozoids swim down to fertilise the vule. The fertilised ovule develops into a yellowish seed with foetid odour.

CONIFERALES

9. The conifers or cone-bearing plants have many livin
species although they too, like the club mosses and ferns, hav
fossil representatives.

10. Habitat. Conifers are found mainly in colder region
such as Canada, Russia and Scandinavia where they tend t
grow on poor soil. They are said to occupy a greater land are
than any other group of plants.

11. Size, shape and structure. Some of the largest organism
in the world, the sequoia or Californian redwoods, are conifer
and grow to over 100 metres. Most conifers are trees but ther
are some low-growing shrubs but no herbaceous plants. Th
timber produced by conifers is fairly dense and massive, unlik
the loose soft wood of the cycads. The leaves are usuall
needle- or scale-shaped with the exception of the fairly large
leaved araucaria or monkey puzzle tree. They are all, how
ever, leathery and this fact, along with modifications such a
small surface area and stomata in pits, helps to cut down th
loss of water. Conifers are usually evergreens although larch i
deciduous.

12. Reproduction. Most conifers are monoecious with th
male and female cones being borne on the same tree. The mal
cones are small and ephemeral, lasting only until the pollen i
shed and then shrivelling and dropping. The female cones
on the other hand are larger and woody and in many specie
are retained on the trees for three years. Large cones ma
reach 60 cm in length. In the juniper and yew, the female cone
are highly modified, being like berries. Common conifers ar
pines, spruces, firs, larches and cedars. Scots pine (*Pinu
sylvestris*) is common in Britain.

PINUS SYLVESTRIS (*SCOTS PINE*)

13. Habitat. The Scots pine is a native of Britain and i
found naturally in Scotland forming rather open forests on shal
low soils. Isolated trees on wind-swept rocky knolls are als
common and it is often found in areas where the water supply i

rather limited. The Scots pine has been widely planted throughout Britain for its timber.

14. Size, shape and structure. The mature tree may reach 30 metres in height. It has a straight tapering trunk with whorls of branches, most of which in an old tree form an umbrella-like head at the top of the trunk. The branches are covered with small brown *scale leaves* which provide some protection and reduce water loss. In the axils of these scale leaves are found *dwarf shoots* of limited growth which bear a pair of needle-like leaves. These needles are about 5 cm long and 2 mm across. They are semi-circular in cross section and therefore have a small surface area. Their stomata are set below the leaf surface in pits. Both of these characteristics help to cut down the loss of water.

The needle leaves are not shed altogether at the end of the growing season; they persist for a few years before being shed along with the dwarf shoot which bears them. The tree is therefore never without leaves and is said to be *evergreen*.

The structures of the tissues and their arrangement in the trunk and roots are very similar to those described for the dicotyledenous angiosperm (*see* VI, **11–18**) except that the xylem is composed chiefly of tracheids and there are no *companion cells* in the *phloem* tissue.

15. Nutrition. The Scots pine is a photosynthetic autotroph. It obtains water and mineral salts from the soil by way of its root system.

NOTE: Many gymnosperms have *mycorrhiza*. These are fungi which live in very close proximity to the roots and partially infect them. The fungi help to provide the trees with mineral elements and appear to obtain food from the trees in return.

16. Reproduction. The Scots pine produces both male and female cones.

(*a*) *The male cones* are produced at the base of the youngest shoots in the *axil* of a scale leaf. They are produced in groups. Each cone is about 1 cm long and has a central axis to which are attached flattened *microsporophylls*. These have two *microsporangia* on their undersides. The microsporangia are protected by the lower edge of the microsporophyll curling down over them and by the curling up of the upper edge of the microsporophyll below.

Within the microsporangia are found the *microspore mother cells* which produce the haploid *microspore* or *pollen grains*. The microspores have two air sacs in the walls to assist their dispersal, which takes place with the aid of the

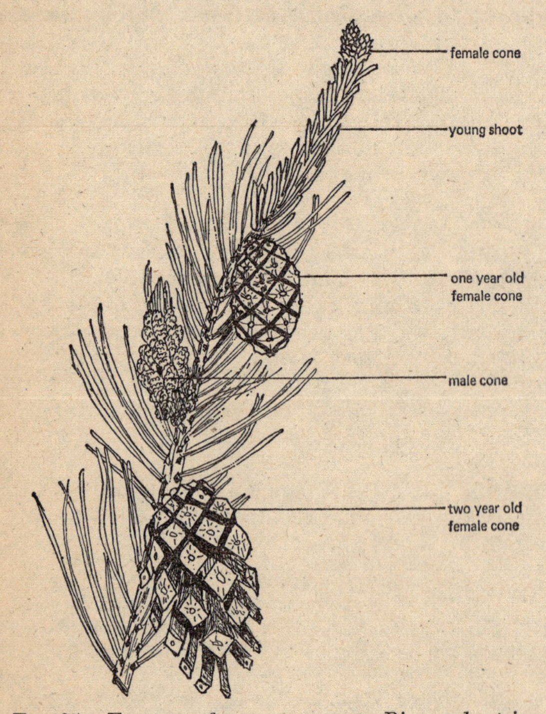

FIG. 25.—Features of a gymnosperm—*Pinus sylvestris.*

wind in May (in Britain). At the time of pollination the microspores or pollen grains consist of two cells, a *prothallial cell* and a *tube cell.*

(*b*) *The female cones* are formed at the very ends of the new shoots but later they are found laterally as the shoots continue to grow. Female cones are about 1 cm in length when first produced and reddish in colour. They consist of a central axis to which are attached *bract scales* and *ovuliferous scales*. The ovuliferous scale is wider and thicker at the edges, ending in a rhomboidal plate which fits neatly against the ends of neighbouring scales.

At the top of an ovuliferous scale are found two developing *ovules*. At first a group of cells forms a *nucellus*. Around the nucellus, two covering layers grow up to form the *integuments* and there is a small gap in the covering called the *micropyle*. Within the nucellus, a single large cell is eventually formed which is called the *embryo sac* or megaspore.

17. Pollination. The cone is now pollinated, the pollen grains or microsporangia being carried in between the ovuliferous scales. These scales have been separated by the inrolling of the bract scales. The nucellus exudes a drop of mucilage from the micropyle and the pollen grain is trapped in it. This drop gradually dries and the pollen grains are drawn in through the micropyle to lie against the nucellus. After pollination, the ovuliferous scales close up again and the cone grows to 4–5 cm in length and becomes green. Development of the ovules also takes place at this time. Repeated divisions of the nucleus of the embryo sac result in a mass of prothallial tissue called the *endosperm*. From this tissue, three archegonia are usually produced. These archegonia have a ventre with an *oosphere* and a *neck* with associated cells. Only one oosphere develops to maturity.

18. Fertilisation. About 11 months after pollination, the pollen grain germinates and grows through the nucellus. The tube cell, after a number of divisions, produces *two male cells* one of which fertilises the oosphere to form a diploid oospore. The fertilised ovule, with its endosperm and oospore undergoes further development to produce a *seed*. The oospore divides and grows to produce an *embryo* with a *radicle* and a *plumule* which is surrounded by a small amount of endosperm, all other tissues within the ovule having been lost. The integuments develop into a gritty-walled seed coat or *testa* which has a membraneous wing to aid dispersal. By the third year the cone has become woody and the ovuliferous scales open again to disperse the seeds.

19. Economic importance.

(*a*) *Cycas revoluta*, the sago palm, has the central pith of the stem packed with starch grains from which the sago of puddings is prepared.

(*b*) *Ginkgo biloba* is grown as a specimen tree in large gardens, especially for its rich autumn colouring.

(*c*) *Pines, firs, spruces and larches* are grown for their timber used in box-making, for poles, in the building industry and for pulping to make paper.

(*d*) *Red cedar* is grown for its timber which is resistant to rots and widely used in constructing glasshouses and bungalows.

(*e*) *Larch timber* of a high grade is used in boat building.

(*f*) *The wood of the yew tree*, formerly used for making long bows, is now used for producing turned ornaments.

PROGRESS TEST 5

1. What are the characteristics of the spermatophytes? **(1)**
2. How are the gymnosperms classified? **(2)**
3. Describe the maidenhair tree, *Ginkgo biloba*. **(7)**
4. What is the most common habitat of the conifers? **(10)**
5. What are mycorrhiza? **(15)**
6. Describe the male and female cones of *Pinus sylvestris*. **(16)**
7. What is the economic importance of the gymnosperms? **(19)**

CHAPTER VI

FLOWERING PLANTS

1. Introduction. The *angiosperms* or flowering plants ppeared relatively recently in the earth's history, the earliest ossils so far found being about 135 000 000 years old. They nay have evolved before that but were unlikely to have been videspread. Plants colonised dry land about 600 000 000 ears ago so that land plants had been evolving for at least 50 000 000 before flowering plants. Even so, the flowering lants now dominate any part of the world where there is egetation, with the exception of cold regions with poor soils which are still the preserve of the conifers. Angiosperms also rovide most of man's food and many of his raw materials.

There are over 250 000 known species of flowering plants of which about 1 500 are found in Britain. The angiosperms are livided into:

(*a*) *monocotyledons*, which include the grasses and cereals;

(*b*) *dicotyledons*, including most shrubs, trees and many common flowers such as roses, dandelions and buttercups.

2. Habitat. Flowering plants colonise almost every habitat vailable to higher plants. They are the most numerous plants n terms of biomass (mass or amount of living tissue) and their uccess is attributed to a number of factors.

(*a*) *Variability in structure.* They show great variability in structure, morphology and physiology and this allows different angiosperms to colonise widely differing habitats.

(*b*) *Genetic flexibility.* The group has considerable genetic flexibility giving greater variation; therefore angiosperms can respond quickly to changes in the environment and produce new species adapted to survive.

(*c*) *Efficient pollination.* The formation of flowers allows for efficient pollination by insects and other means.

(*d*) *Production of large numbers of seeds.* Many angiosperms have a fast rate of growth and a short life cycle and can

produce large numbers of seeds. This enables the angio
sperms to be very efficient at colonising bare ground such a
that caused by the death of old trees in a forest or even by
fire.

3. Size, shape and structure. The angiosperms exhibit suc
diversity of forms that it is impossible to survey them ade
quately without a lengthy treatment. The aim here will be t
consider general features of a dicotyledonous flowering plan
while making some mention of major variations, especially i
the monocotyledons.

Before we discuss general features of structure, it is appro
priate to consider the types of tissues which make up th
angiosperm body, many of which have already been mentione
in earlier descriptions of lower plants.

PLANT TISSUES

4. Cells and tissues. In Chapter I we looked at a generalise
plant cell. In fact it would be very difficult to find such a ce
in a plant. As with most other living things, a certain amoun
of *organisation* and *division of labour* has taken place within th
plant. Cells have been modified so that they can fulfil differen
functions within the plant more easily. This modificatio
process takes place during growth after cell division and i
termed *differentiation*.

After differentiation, cells of a similar type are usually foun
together forming a *tissue*. The main types of tissue and th
fuctions they fulfil are described below.

5. Meristematic tissues.

(*a*) *Structure*. They are composed of fairly small, more o
less box-shaped cells which have a large nucleus and are ful
of cytoplasm and do not have any vacuoles (*see* Fig. 26).

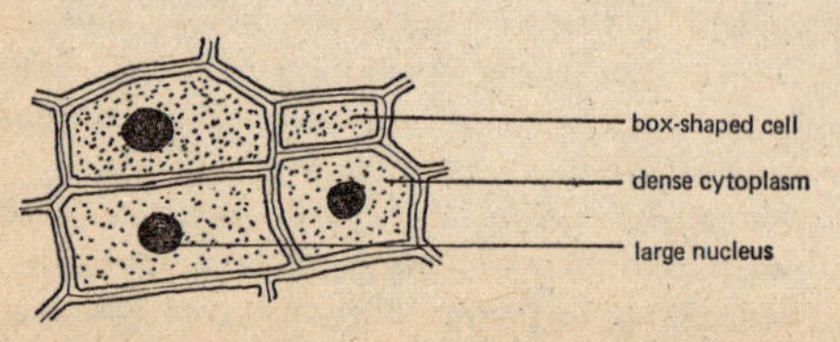

FIG. 26.—Structure of meristematic tissue.

(*b*) *Function.* This tissue provides for growth, either in length or width through the cells dividing by mitosis.

(*c*) *Site in plants.* Shoot and root tips and around the circumference of trees to provide for increases in girth.

3. Parenchyma.

(*a*) *Structure.* The cells of parenchyma tissue are larger than cells of meristematic tissue because they have enlarged and become vacuolated. They are sub-spherical or brick-shaped and some may contain chloroplasts (*see* Fig. 27).

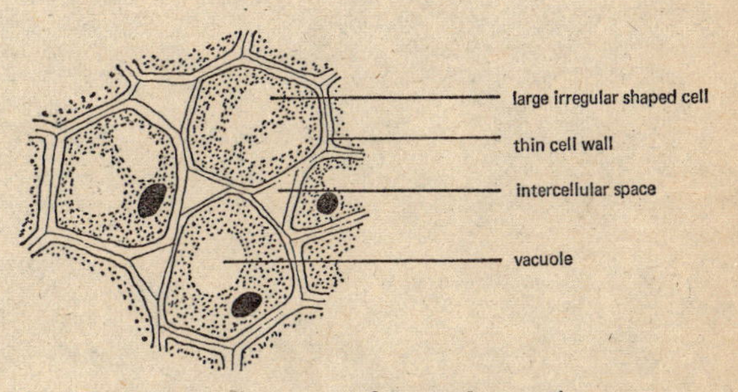

FIG. 27.—Structure of parenchyma tissue.

(*b*) *Function.* Parenchyma tissue with sub-spherical cells is often found packing the areas between other tissues. As well as filling in, the cells may also store food reserves. Other sub-spherical and brick-shaped cells contain chloroplasts and are therefore able to photosynthesise and so produce food for the plant. Both types of tissues are mainly responsible for supporting soft fleshy plants, being able to do this because the cell vacuole becomes full of sap. This in turn causes pressure on the cell wall and the cell becomes stiffened. This is called *turgor pressure* and the plant is said to be turgid. When the cells lose water they become soft and floppy—the plant wilts and becomes flaccid.

(*c*) *Site in plants.* Packing and storage parenchyma are found in the centres of herbaceous stems and photosynthetic parenchyma is found in leaves and other green structures in plants.

7. Collenchyma.

(*a*) *Structure.* The cells are somewhat like parenchyma ir that they have cytoplasm, vacuoles and may even have chloroplasts. They may, however, be more elongated and have thicker cellulose walls. This extra thickening may be uniform on all sides or it may just be confined to the corners (*see* Fig. 28).

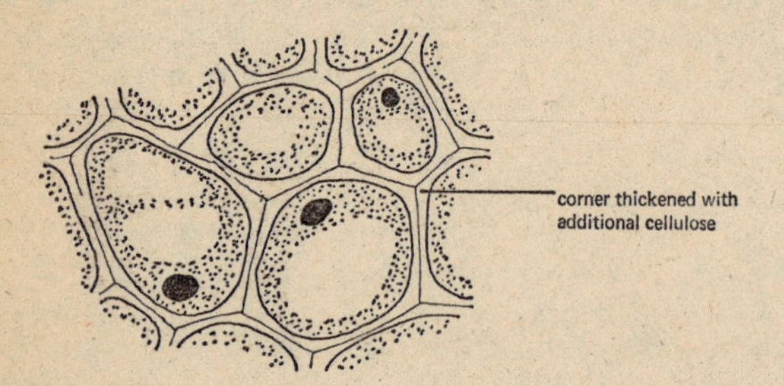

Fig. 28.—Structure of collenchyma tissue.

(*b*) *Function.* It provides support and to a much lesser extent, areas for photosynthesis.

(*c*) *Site in plant.* It is often found in a ring or in patches around the periphery of the stem.

8. Sclerenchyma.

(*a*) *Structure.* There are two types of sclerenchyma:

(*i*) *fibres* are elongated with wedge-shaped interlocking ends;

(*ii*) *sclereids,* sometimes termed "stone cells", are irregularly shaped.

Both types of sclerenchyma have fairly thick lignified cell walls and do not have any contents, i.e. no cytoplasm or nucleus. The protoplast dies after the cells become fully grown and the walls become lignified (*see* Fig. 29).

(*b*) *Function.* Sclerenchyma provides protection for more delicate tissues like phloem and also provides support.

(*c*) *Site in plant.* It is found in patches outside the phloem (*see* **16** below) in some species or in rings around the

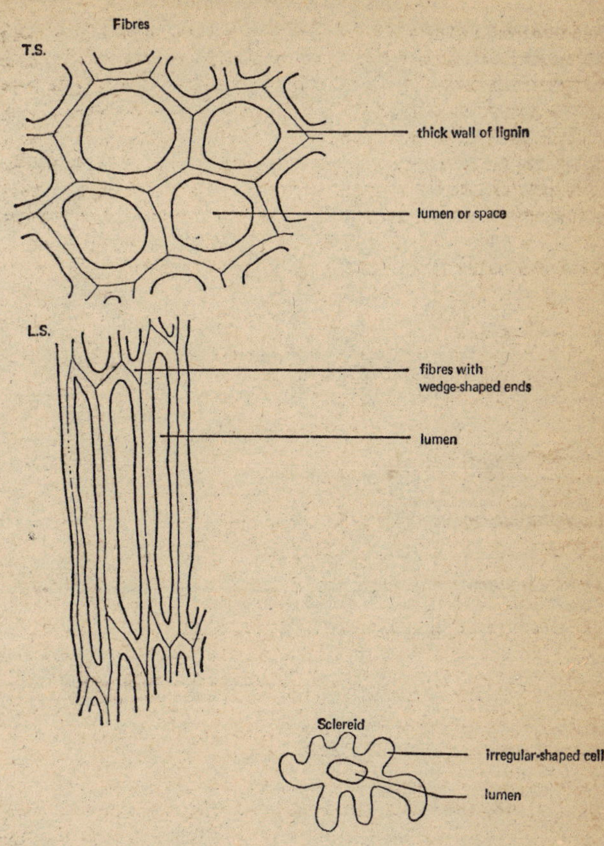

FIG. 29.—Structure of sclerenchyma tissue. T.S., transverse section; L.S., longitudinal section.

stem. The sclereids are the cells which give the gritty texture to unripe pears. They also form the shells of nuts.

9. Phloem.

(*a*) *Structure.* Phloem also has two types of cells, but here the cells are found together and are interdependent. The two types are *sieve tubes* and *companion cells.*

(*i*) *Sieve tubes* are elongated and arranged end to end. The walls are unthickened and the end walls have perforations in them. They are therefore called *sieve plates*. The cytoplasm lies loosely in the centre of the cell, surrounded by cell sap. At the ends of the cell, however, the cytoplasm fans

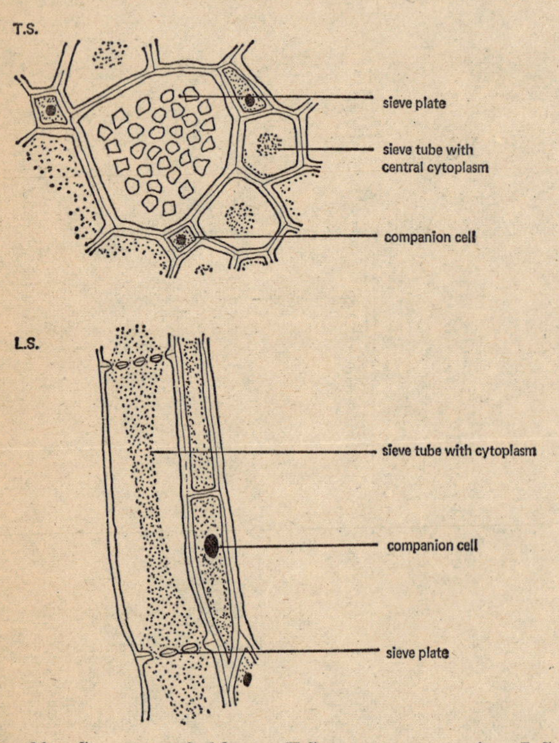

FIG. 30.—Structure of phloem. T.S., transverse section; L.S., longitudinal section.

out and strands pass through from one cell to the next. There are no nuclei in the sieve tubes (*see* Fig. 30).

(*ii*) *Companion cells* are small cells which lie alongside the sieve tubes. They have dense protoplasm.

(*b*) *Function.* The function of the sieve tubes is to provide a route for the transport or *translocation* of the products of photosynthesis. Companion cells, with their nuclei, seem to control the translocation in the sieve tubes.

(*c*) *Site in the plant.* In soft herbaceous stems, the phloem is found on the outside of bundles of xylem tissue, but in woody stems it forms a thin layer just beneath the woody bark.

10. Xylem.

(*a*) *Structure.* The two main types of cells in xylem tissue are *tracheids* and *xylem vessels* or *elements.*

(*i*) *Tracheids* are very long cells, often up to 5 mm long, with tapered ends. The walls are thickened with lignin. This is laid down as a complete cover all over except for

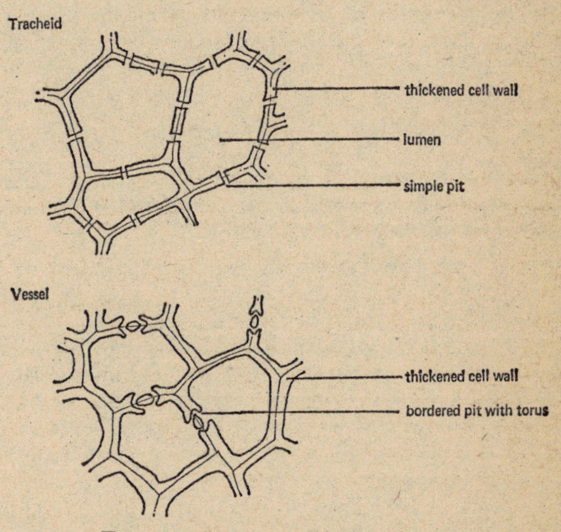

FIG. 31.—Structure of xylem tissue.

areas of interconnecting pits where the lignin is discontinuous. They have no protoplasm and are filled with water and mineral salt solutions. As with sclerenchyma, the protoplasm dies as the cells mature and become lignified (*see* Fig. 31).

(*ii*) *Xylem vessels or elements.* Here the cells are also long but the end walls have also disappeared for great lengths so that continuous tubes of cells up to 3 metres in length have been recorded. The early xylem to be formed, the *primary xylem*, loses its protoplasm and the walls become thickened with lignin which may be laid down in rings, spirals or patches or form a complete cover except for the pit areas. In *secondary xylem* again there is no protoplasm and the walls have a thick layer of lignin except for pit areas. The pits may be complex *bordered pits* in which a rim is found around the area of the pit and a plug or *torus* fills the middle. Like tracheids, vessels contain water and mineral salts.

(*b*) *Function.* The function of xylem is to transport water and mineral salts from the roots to other parts of the plant. Scattered within the xylem are parenchymatous cells which help to control the flow.

(*c*) *Site in plants.* Found in small patches on the inside of the phloem towards the centre of stems, making up all the woody central parts of tree trunks and branches, and found in conducting tissues in roots and leaves.

THE STRUCTURE OF THE FLOWERING PLANTS

11. General features. A generalised diagram or "blueprint" of a flowering plant is given in Figure 32 and reference may be made to the main parts of the plant.

(*a*) *The roots* provide anchorage in the soil for the plant, they obtain water and mineral salts from the soil, and they act as areas for storage of materials.

(*b*) *The stem* provides structural support for leaves and flowers, is often green and capable of photosynthesis, has conducting tissue allowing an exchange of materials between different parts of the plant, may act as a food store and may also act as a storage place for waste materials (e.g. in *bark*).

(*c*) *The leaves* are primarily organs of photosynthesis; their structure and shape relates to this, being long and thin and orientated towards the light.

(*d*) *Flowers* are the sites of sexual reproduction in plants and are usually borne on aerial shoots for two reasons:

(*i*) *to facilitate cross-fertilisation with other plants;*
(*ii*) *to give greater potential for seed dispersal.*

12. The vascular system. In most flowering plants there is a continuous vascular system which permeates all the organs of the plant and provides the means for transporting components of the plant from their sites of entry or synthesis to their sites of use or storage. Gases such as oxygen and carbon dioxide are normally available in close proximity to any part of the plant (even roots in the soil) and are not transported to any

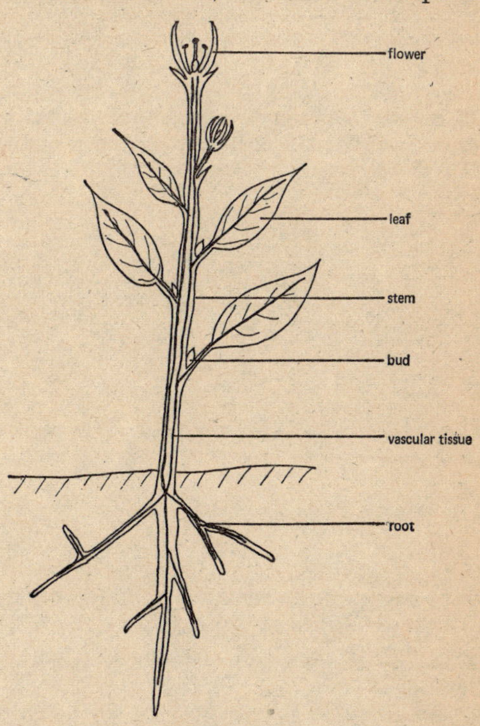

FIG. 32.—The general features of a flowering plant.

extent through vascular tissues, but rather through air spaces between cells when necessary.

The conductive cells in vascular tissues are frequently associated with structural cells such as sclerenchyma, although the conductive cells themselves have considerable strength. The various cells make up *vascular bundles* and these are important in maintaining the structure of a plant. The vascular bundles of stem and root vary in their arrangement because of the different forces to which these organs are exposed.

Stems and shoots are mainly subject to bending forces, either resulting from the weight of leaves and flowers they are supporting or from the effects of wind and uneven growth. Their vascular bundles are arranged like a tube, the best structure to meet such stresses. Roots, however, anchor a plant and therefore have to withstand pulling. In a heterogeneous medium like soil, a tube of vascular bundles would tend to be crushed and root vascular bundles therefore tend to form a central core with the change from root core to stem tube taking place at about ground level (*see* Fig. 33).

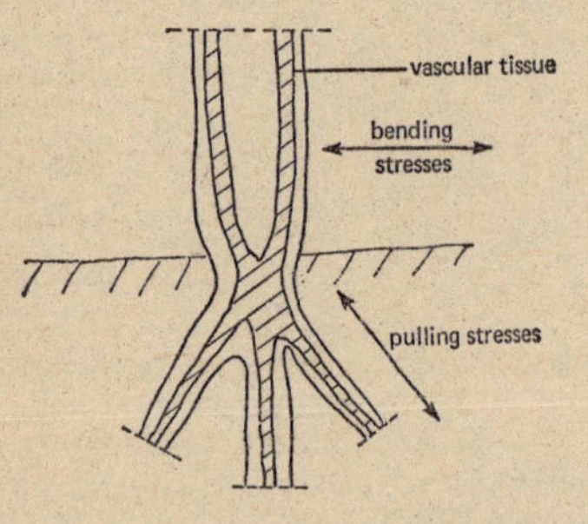

Fig. 33.—The arrangement of vascular tissues in the root and shoot of a flowering plant.

13. The structure of the root. There are two kinds of root system in a flowering plant:

(*a*) *The tap root system* in which a primary root dominates the root system, with branches coming off it;

(*b*) *The fibrous root system* in which all the roots are thin and diffuse.

In both systems, there are four fairly well-defined zones in a root.

(*a*) *A root cap* of cells which tend to get sloughed off and repeatedly replaced as the root pushes through the soil. The root cap protects a small *apical meristem*, a region of meristematic tissue where constantly dividing cells provide the basic "cell stock" for the root.

(*b*) *The growing region* where the new young cells produced in the apical meristem are elongating and starting to differentiate into tissues.

(*c*) *The absorptive region* where tissues are fairly well differentiated and in which the outer epidermal cells are

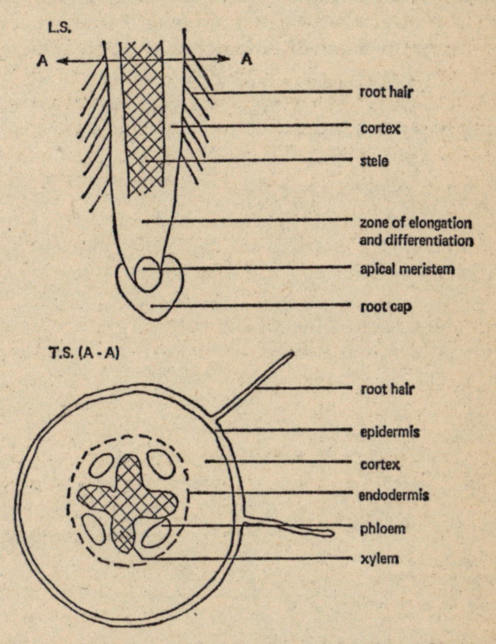

FIG. 34.—Longitudinal and transverse sections through a young root.

often elongated into *root hairs* which extend into the soil and take up water and mineral salts.

(*d*) *The mature root* which does not have root hairs but does take up some water through the epidermal cells. Its functions include anchoring the plant, transporting materials and storing food.

14. The anatomy of a young root. Figure 34 shows a longitudinal section of the tip of a root and a transverse section through the absorptive region. An outer piliferous layer of water-absorptive cells termed the epidermis produce root hairs as extensions of individual cells. Under this layer is an exodermis which can be suberised to form a protective coat in older roots. Between this and the core of conductive tissue is the *cortex* made up of starch-filled parenchymatous cells.

The conductive tissue consists either of a central xylem core of vessels or several clusters of vessels and some parenchyma, surrounded by a number of clusters of phloem cells.

NOTE: Figure 34 shows a *tetrarch* root with four phloem clusters. Roots may also have two (*diarch*), three (*triarch*), several or many (*polyarch*) clusters. Polyarch roots tend to have a central pith of parenchyma cells.

Between the xylem and phloem tissues are actively growing cambium cells. Surrounding this core of conductive tissue, the *stele*, and separating it from the cortex is the *endodermis*, a layer of water-proofed cells which controls water movement.

15. Secondary thickening in the root. This occurs as a root matures and gives it increased abilities to withstand physical stresses and also to transport materials. It consists of layers

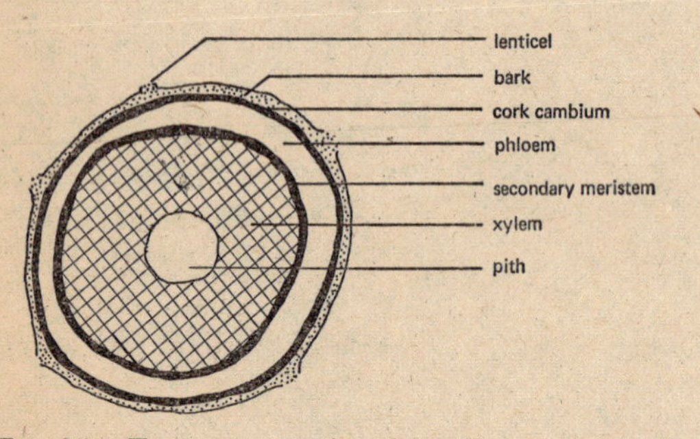

FIG. 35.—Transverse section through a mature root.

of additional xylem and phloem tissues laid down by the *cambium tissue* which forms between the initial phloem and xylem bundles and is termed a *secondary meristem*, being secondary to the primary apical meristem. This meristem forms additional phloem on its outer side and xylem on its inner side so that the root with secondary thickening has its oldest xylem tissue near the centre but its oldest phloem on the outside.

At the same time as the secondary xylem and phloem are formed, cells on the outside of the *stele* form a protective layer of bark which, incidentally, isolates the outer cells of the cortex which slough off. The mature root (Fig. 35) therefore consists largely of vascular tissue surrounded by bark with, perhaps, a small central zone of pith cells.

16. The structure of the stem. A stem of an annual flowering plant has an outer layer of epidermis cells which have a coating of a waxy cuticle resistant to water loss. Below this is the cortex, consisting of:

(*a*) *an outer band* of thick-walled photosynthetic collenchyma;

(*b*) *an inner band* of thin-walled parenchyma, mainly used for storing food.

As shown in Fig. 36 the vascular bundles of a typical dicotyledenous flowering plant form an incomplete tube around a central zone of parenchyma, the pith. This tissue, the parenchyma, extends outwards to the cortex in the form of *medullary rays* in between the vascular bundles and, like the parenchyma of the cortex, is an important storage tissue.

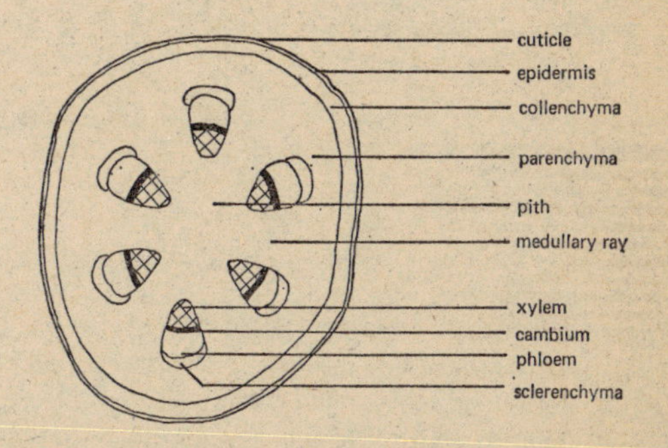

FIG. 36.—Transverse section through a young shoot.

The vascular bundles consist of xylem tissue nearest the centre separated from the phloem by a layer of cambium, and on the outside of the phloem is a zone of thick sclerenchyma giving protection and structural support (*see* Fig. 36).

17. Secondary thickening in the stem. The stem of a woody perennial plant grows in height and girth each year and needs additional support and conductive tissue. This is acquired through secondary thickening, the production of layers of secondary xylem and phloem within the stem. Initially the cambium separating xylem and phloem in the vascular bundles extends through the medullary rays to form a complete ring of cambium, a secondary meristem. It then forms a ring of secondary xylem on the inside and secondary phloem on the

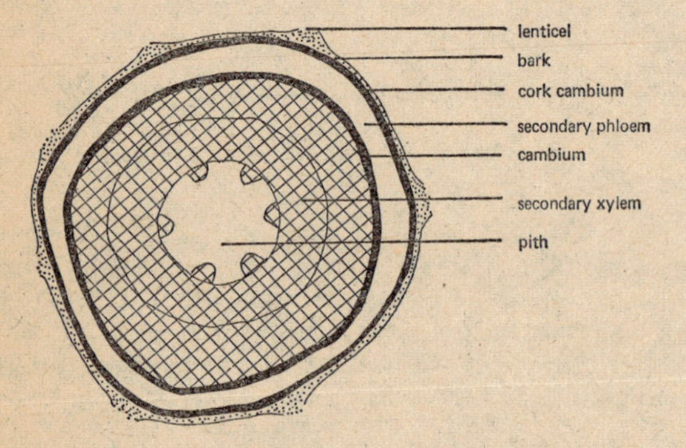

Fig. 37.—Transverse section through a 3-year-old stem with 2 years' growth of secondary thickening.

outside, further rings being formed in each growing season. Figure 37 shows a 3-year-old stem with 2 years of secondary thickening.

18. Bark. In woody plants, as the secondary xylem and phloem increase, the cortex and epidermis start to get shed and another secondary meristem, the *cork cambium*, forms within the inner layers of the cortex. This replaces the outer cortex with layers of corky cells impregnated with *suberin*, forming a protective covering for the stem phloem and also preventing water loss. This covering of bark is also fairly gas-tight, and to ensure gaseous exchange between tissues inside the stem and the surrounding atmosphere, *lenticels*, or channels of loose cork cells are formed in the bark. The position of the lenticels corresponds to that of the stomata in the disrupted epidermis.

19. Growth of the stem. In most angiosperms there is an apical meristem in the stem which, like that of the root, just behind the root cap, is responsible for forming new stem tissues. But because the stem forms leaves and flowers and even roots, the region of the shoot apical meristem is far more complex than that of the root. On its activities depend the formation of leaves as well as *axillary buds*, between leaves and stem, which in turn can form leaves or flowers. In many flowers, the apical meristem of the original main shoot will ultimately form a flower bud. This is particularly common in monocotyledons.

20. The anatomy of the leaf. The main function of the leaf is photosynthesis, and apart from regions of active growth, leaves

FIG. 38.—Arrangement of veins, or vascular bundles, in a sycamore leaf.

are the most chemically active components of a plant. They are normally thin and horizontally orientated to absorb the maximum amount of sunlight, and leaves of some plants will even change orientation during the course of the day.

Because of its large surface area, a leaf requires good conducting tissues to bring water and mineral salts from the roots and it also has to have adequate rigidity. Leaf veins or vascular bundles serve both functions and Fig. 38 shows the spread of veins in a sycamore leaf.

Structural support is also provided by turgid cells within the leaf and one of the first symptoms of water loss in leaves is wilting, brought about by loss of turgor.

Figure 39 shows a transverse section of a typical leaf of a

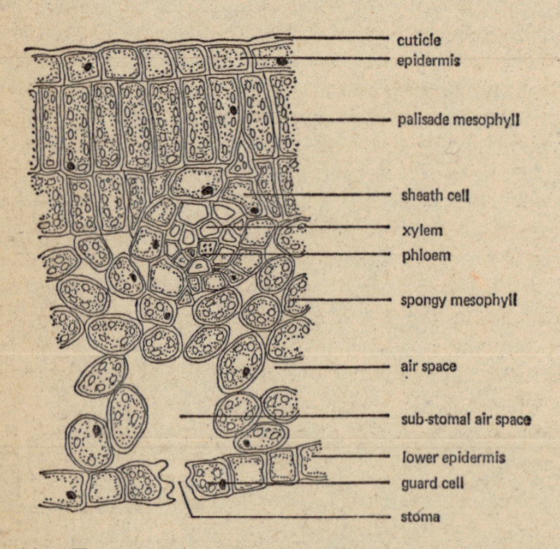

FIG. 39.—Transverse section through part of a typical leaf.

dicotyledonous flowering plant. There are three main tissue components, mesophyll, vascular bundles and the epidermis.

(*a*) *Mesophyll* tissue makes up the bulk of the leaf and consists of two forms of parenchyma.

(*i*) *Palisade mesophyll cells.* The cells near the upper leaf surface are tightly packed and elongated at right angles to the surface. These are *palisade mesophyll* cells, usually two or three layers thick and with large numbers of chloroplasts.

(*ii*) *Spongy mesophyll cells.* Extending from the palisade mesophyll to the lower epidermis are *spongy mesophyll* cells. As the name implies, these cells are not packed together but have plenty of air spaces between them which are in contact with the atmosphere through stomata (*see* (*c*) below).

Both forms of mesophyll are active in photosynthesis but the palisade mesophyll cells near the upper surface have up to three times as many chloroplasts as the spongy mesophyll.

NOTE: Parenchyma cells containing chloroplasts occur in stems and unripe fruit as well as leaves. Wherever it occurs, such parenchyma is often termed *chlorenchyma*.

(*b*) *Vascular bundles*. These are the leaf veins and include xylem and phloem elements as well as sclerenchyma fibres providing structural support. The main bundles, continuous with *petiole* and stem vascular tissue, branch repeatedly to form a network extending throughout the leaf. Each vascular bundle is usually midway between the upper and lower leaf surfaces and is surrounded by a sheath of firm parenchyma cells. These have few chloroplasts and are elongated parallel to the line of the vascular bundle. This *bundle sheath* is in direct contact with surrounding mesophyll cells and material passing between vascular tissues and the mesophyll cells pass through it. The bundle sheath cells may extend vertically from the vascular bundle to the upper and lower epidermis.

(*c*) *Epidermis*. This consists of a single layer of small cells, usually without chloroplasts and with the outer surface coated with a protective layer of waxy cuticle. The epidermis is continuous around the leaf but the upper epidermis normally has a heavier cuticle lining than the lower epidermis, although water loss through each layer is minimal.

The lower epidermis has large numbers of pores termed stomata (sing. stoma). These lens-shaped pores are really spaces between cells but each is surrounded by two specialised kidney-shaped cells of the epidermis which have chloroplasts and are called guard cells (*see* Fig. 40). The walls of guard cells are differentially thickened so that changes in turgor pressure alter their shape and open and close the stoma. The factor indirectly causing the change in guard cell turgor and hence shape is light, so that stomata open in the day and close at night. The leaf is therefore able to exchange gases and water vapour with the surrounding atmosphere when it is actively engaged in photosynthesis during the day.

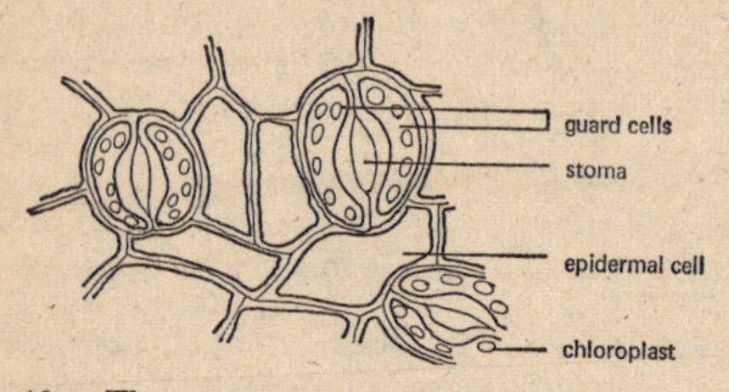

FIG. 40.—The arrangement of cells around stomata.

21. The petiole is the stalk of the leaf. It usually has one vascular bundle which connects with a stem vascular bundle and continues into the leaf to form the main leaf vein. The petiole vascular bundle is normally central and the petiole has stomata and is usually actively photosynthetic.

22. Abscission. Many angiosperms are deciduous, shedding their leaves before adverse cold or dry seasons. Leaf fall results from the formation of an *abscission layer* of parenchyma cells extending across the base of the petiole. At the end of the growing season, the leaf breaks at this point, and a layer of cork is produced to protect the stem.

23. Modifications of leaves include:

(*a*) *spines*, usually serving a protective purpose as in the gooseberry;
(*b*) *scales*, which may protect buds or may be thick and store food, as in bulbs;
(*c*) *floral leaves*—it is thought that flowers are largely modified leaves;
(*d*) *tendrils*, in climbing plants like the sweet pea;
(*e*) *cotyledons*, the first leaves produced by germinating seeds and usually very fleshy.

ANATOMY OF THE MONOCOTYLEDONS

24. Monocotyledon features. The descriptions given above for roots, stems and leaves apply to dicotyledonous flowering plants. Some of the major ways in which monocotyledons differ from this are as follows.

(*a*) *The root.*

(*i*) *They are polyarch* with many xylem and phloem groups and with a central pith.
(*ii*) *They often have a much thicker endodermis.*

(*b*) *The stem.*

(*i*) *There are usually far more vascular bundles.* They are distributed less regularly (although the xylem is always towards the centre) and there is no clear differentiation into cortex and pith.
(*ii*) *There is no cambium within the vascular bundles* and therefore no secondary thickening, although each bundle may be surrounded by sclerenchyma cells.

(*c*) *The leaf.*

(*i*) *Each seed produces a single cotyledon,* as the name implies, whereas in dicotyledons there are two.
(*ii*) *The leaves are usually long and thin* with parallel rather than branching veins and are less rigid than in most dicotyledons.

ASEXUAL REPRODUCTION IN FLOWERING PLANTS

25. Forms of asexual reproduction. There are many methods of asexual or *vegetative reproduction* in flowering plants and the border-line between true asexual reproduction and the survival of a storage organ through adverse conditions is ill-defined. Vegetative reproduction is most common in herbacious and woody perennials and the methods include:

(*a*) *production of runners*, as in the wild strawberry, where modified stems grow along the ground and eventually root and produce new plants;
(*b*) *production of underground stems, stolons and rhizomes*, which grow through the soil and send up new shoots;
(*c*) *the natural rooting of ordinary aerial shoots* when they bend over and touch the ground, as in the blackberry.

Many plants survive dry or cold seasons by producing storage organs which lie dormant, the remainder of the plant decaying. Many plants may produce a large number of such storage organs with each organ able to form a new plant in the following growing season (e.g. the potato).

26. Advantages of vegetative reproduction.

(*a*) *In wild plants.*

(*i*) *It can be a very rapid process.*
(*ii*) *A new plant can be formed while still relying on food from the parent plant.*

(*b*) *In cultivated plants*—advantages for man.

(*i*) *The organs of vegetative reproduction may be good food sources,* e.g. potato, cassava and yams (*see* X).
(*ii*) *Propagation is often easier and quicker than with seeds.*
(*iii*) *Vegetative reproduction ensures that the offspring is identical to the parent.* It is known as a *clone.*

SEXUAL REPRODUCTION IN FLOWERING PLANTS

27. The flower. Although gymnosperm sporophylls may be called flowers, the term is usually reserved for angiosperms. Being members, like the gymnosperms, of the spermatophyta, the adult plants in the angiosperms are the sporophyte generation, with the gametophyte generation being very restricted and entirely dependent on the sporophyte. The megaspore still forms an ovule along with protective integuments, but in the angiosperms, there is an additional protective tissue, the *ovary.*

28. The structure of the flower. There is considerable diversity in floral form and the following description will concentrate on general features. The flower is a modified shoot with the micro- and macrosporophylls called *stamens* and *carpels* normally occurring in whorls (the *androecium* and the *gynaecium* respectively) rather than in spirals as in the gymnosperms. They usually occur on the same flower, which is therefore termed *hermaphrodite.* The terminal shoot tissue bearing the stamens and carpels is termed the *receptacle* and the whole shoot axis is the *peduncle.* Below the stamens and carpels it bears two whorls of specialised leaves, the *perianth* of sepals (*see* Fig. 41).

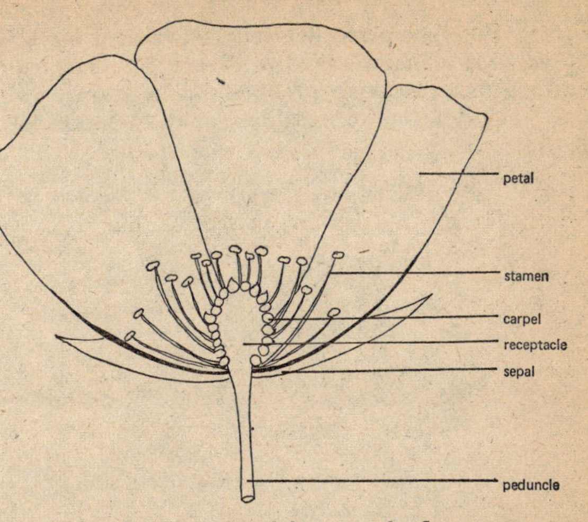

FIG. 41.—The general features of a flower.

29. The perianth consists of:

(*a*) *an outer calyx* made up of a number of fairly tough *sepals* which protect the immature flower before it opens;
(*b*) *the corolla* made up of *petals*, often brightly coloured in insect pollinated flowers but much reduced and drab in wind pollinated flowers. Additional insect attractants may be present and include scents and nectar, the latter being a sugar-rich liquid produced at the base of the petals in *nectaries*.

The arrangement of petals may be:

(*a*) *zygomorphic*—bilaterally symmetrical;
(*b*) *actinomorphic*—radially symmetrical.

Petals and sepals may be more or less fused to form tubes.

30. The androecium is the whorl of stamens, which are made up of:

(*a*) *a filament* or slender flexible stalk;
(*b*) *an anther* or microsporangium with four *pollen sacs*.

After initial development, each pollen sac contains *pollen mother cells* (i.e. spore mother cells) nourished by a layer of *tapetal cells* which line the sac (*see* Fig. 42). The pollen mother cells divide meiotically to produce pollen grains, the microspores, which develop an irregular sculptured exterior. Each

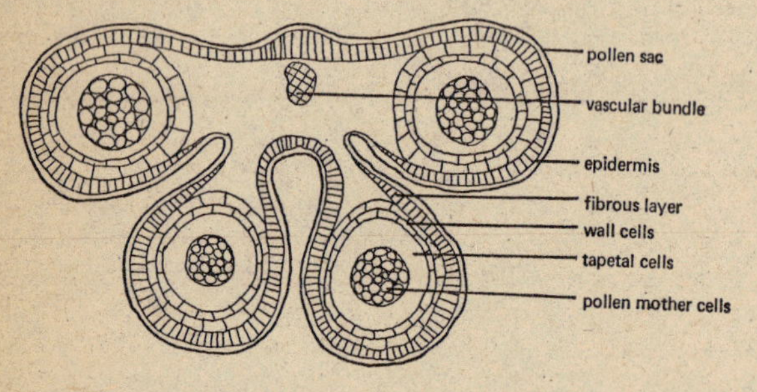

FIG. 42.—Transverse section through a stamen.

pollen cell nucleus divides mitotically to produce a *tube nucleus* and a *generative nucleus* and the pollen grains are then liberated by *dehiscence* or rupture of each pollen sac.

31. The gynaecium is borne terminal to the androecium as a whorl of carpels on the end of the receptacle. Each carpel consists of a hollow ovary containing one or more *ovules* and extending by way of a slender *style* to form a *stigma*, a receptive surface for pollen grains. Within the ovary each ovule is attached to the ovary wall at a point called the *placenta* by a stalk or *funicle*. The ovule develops initially as a mass of cells or *nucellus* which differentiates to form *inner* and *outer integuments* which grow round the nucellus and entirely enclose it except for a small pore called the *micropyle* at one end. The opposite end of the nucellus to the micropyle is termed the chalaza (*see* Fig. 43).

Within the nucellus, a diploid *embryo mother cell* produces four haploid cells, three of which degenerate and the fourth developing into an *embryo sac*. The nucleus divides mitotically three times to produce eight nuclei, four at each end of the sac. One from either end migrates to the centre and these are called *polar nuclei* but do not form cell walls. The remaining six nuclei form six cells, those at the chalaza end being called *antipodal cells* and those at the micropyle end developing into the *ovum* or egg cell and two non-functional ova called *synergids*. A carpel may have one or several ovules arranged in various patterns.

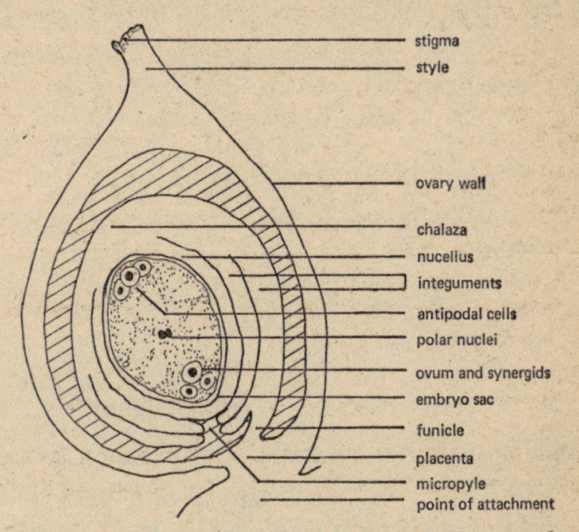

FIG. 43.—Longitudinal section through a carpel.

32. Pollination is the process of transferring pollen to the stigma of a carpel and may involve insects, wind or even water or birds. A pollen grain alighting on a stigma adheres to the sticky surface and germinates to produce a pollen tube growing down through the style towards the ovule, with the tube nucleus occupying a position at the end of the tube. The pollen tube normally enters the ovule through the micropyle.

33. Fertilisation. The generative nucleus of the pollen grain enters the tube and divides mitotically to form two *male nuclei* which migrate to the end of the tube, now grown up against the wall of the embryo sac. Portions of the pollen tube and embryo sac walls degenerate and the two male nuclei enter the embryo sac, one fusing with the female ovum and the other with a *secondary nucleus*, formed by prior fusion of both polar nuclei, to form a *tripolid nucleus*, thus resulting in a *double fertilisation* unique to angiosperms:

ovum + male nucleus = diploid embryo
secondary nucleus + male nucleus = triploid endosperm.

34. Seed formation.

(*a*) *The triploid endosperm* cell divides repeatedly to form a food store for the seed.

(*b*) *The fertilised ovum* divides and one of the resulting cells divides again repeatedly along one plane to form a *suspensor*, a long chain of cells with the *embryo cell* at one end which is pushed up into the endosperm where it develops into the embryo, forming cotyledons, a *plumule* or shoot and a *radicle* covered by a root cap.

The stamens and perianth degenerate and the gynaecium develops with the integuments forming a hard seed coat or *testa* surrounding the embryo and endosperm to produce a *seed*, this being enclosed by a *fruit* or *pericarp* which develops from the ovary wall.

SEEDS AND FRUITS

35. The endosperm is not necessarily retained in the seed.

(*a*) *Endospermic seeds* do retain it as a food store to be used when the seed germinates (e.g. maize).

(*b*) *Non-endospermic seeds* such as sun-flower and broad bean tend to use up the endosperm in forming food stores in the cotyledons.

36. Fruits. Within the angiosperms are many kinds of fruit; some of them include parts of the plant adjacent to the ovary. Fruits are grouped into four major categories.

(*a*) *Indehiscent dry fruits* do not split open, are often hard and usually contain a single seed. They include woody *nuts* such as hazel and leathery *achenes* such as in the buttercup.

(*b*) *Dehiscent dry fruits* usually contain many seeds and split open in various ways to liberate the seeds. They include *legumes* such as peas and beans and *capsules* such as the poppy and foxglove.

(*c*) *Schizocarpic fruits* usually develop from several carpels and, as the name implies, they do not dehisce, but the carpels, each of which contains a seed, split away from each other. An example is the geranium.

(*d*) *Succulent fruits* include soft fleshy *berries* with many seeds like the tomato and grape, as well as *drupes* (the stone-fruits), each containing one seed (e.g. plum). They also include *pseudocarps* or false fruits such as *pomes* (e.g. apple) and *hips* (e.g. rose) where the receptacle, in addition to the ovary, forms part of the fruit.

37. Fruit and seed dispersal. Effective dispersal aids

(*a*) *the colonisation of new ground;*
(*b*) *the reduction of competition* between parent and young plants.

Dispersal may be aided by wind, water, animals and even by mechanical propulsion.

(*a*) *Wind dispersal* usually involves small light seeds or else seeds with wings or feathery plumes.
(*b*) *Water dispersal*, most common in aquatic and coastal plants, involves seeds that float.
(*c*) *Dispersal by animals* involves dispersal on or in an animal. Small seeds may lodge in mud on the feet of animals, other seeds have hooks or barbs which catch on to fur and many other seeds are contained in succulent fruits. These latter are eaten by an animal, pass through the gut and are excreted in the faeces, a very fertile environment for germination!
(*d*) *Mechanical propulsion* exists in some plants where seeds are expelled, often by a hygroscopic mechanism. The 3 mm long seed of *Arceuthobium* (dwarf mistletoe), for example, can be propelled over 20 metres.

38. Seed germination. The general rule in seed germination is that it will not take place until environmental conditions are favourable and this rule can even extend to mechanisms that prevent a seed from germinating *unless* it has experienced a cold or dry season. Other aspects of seed germination may be mentioned.

(*a*) *Water* is a normal requirement for germination and is often taken in, mainly through the micropyle.
(*b*) *The temperature* must be within a certain range, usually a few degrees either side of the optimum range for growth of the seedling, but the range varies greatly in the angiosperms.
(*c*) *Oxygen* is a normal requirement as *aerobic respiration*

using oxygen is far more efficient at releasing the limited energy content of a seed's restricted food store than is *anaerobic respiration* (*see* VII, 32).

(*d*) *Light* is a variable requirement, many seeds being unaffected but some requiring light and others being inhibited by it.

STORAGE ORGANS

39. Storage organs are not necessarily forms of vegetative reproduction, as many plants will utilise food previously stored in these organs to survive a period of adverse environmental conditions. Plants store food in almost every tissue, but storage organs are major modifications of plant parts, usually the stem and leaves.

40. Examples of storage organs.

(*a*) *Rhizomes* are underground stems, often greatly swollen as in the iris.

(*b*) *Tubers* (e.g. the potato) are also underground stems but are branches of the main stem, unlike rhizomes which consist of the main stem itself.

(*c*) *Corms* are also thickened stems which bear thin scale leaves and apical buds, and are usually situated just below ground level, e.g. the crocus.

(*d*) *Bulbs* are short stems with fleshy scale leaves storing the food. The outermost leaves may be thin and protective and there may be buds in leaf axils. Examples are onions, tulips, hyacinths and daffodils.

(*e*) *Swollen tap roots* such as parsnips and carrots.

CLASSIFICATION OF FLOWERING PLANTS

41. The main division of the flowering plants, as already mentioned, is into monocotyledons and dicotyledons. There are several methods of classifying them in great detail, mostly dependent on floral form, especially the arrangement of the petals and gynaecium. One form of classification is summarised below.

(*a*) *Dicotyledons.* Perianth usually made up of sepals and petals, arranged in groups of four or five.

(*i*) *Polypetalae*. Several (usually five) separate petals e.g.
Ranunculaceae (buttercups)
Rosaceae (roses)
Leguminosae (beans and peas).

(*ii*) *Sympetalae*. Petals partly or wholly fused e.g.
Primulaceae (primroses)
Solanaceae (potatoes and tomatoes).

(*iii*) *Apetalae*. No petals e.g.
Salicaceae (willow).

(*b*) *Monocotyledons*. Perianth made up of petals only, usually in threes (*see* **24** above for anatomical differences).

(*i*) *Petaloidae*. Petals present e.g.
Liliaceae (lilies)
Iridaceae (irises).

(*ii*) *Glumiferae*. Petals rudimentary or absent e.g.
Graminae (grasses).

PROGRESS TEST 6

1. When did flowering plants first appear on Earth? (**1**)
2. Why have flowering plants been so successful? (**2**)
3. List the major kinds of plant tissues. (**4–10**)
4. Describe the shape and structure of xylem cells. (**10**)
5. What are the functions of the different parts of a flowering plant? (**11**)
6. Why does the arrangement of the vascular tissue in a root differ from that in a stem? (**12**)
7. What are the four main regions of a young root? (**13**)
8. How do dicotyledon roots differ from monocotyledon roots? (**14, 15, 24**).
9. How does secondary thickening take place in roots and stems? (**15, 17**)
10. What is the function of bark? (**18**)
11. What are the main features of leaf mesophyll tissue? (**20**)
12. How does abscission occur? (**22**)
13. Describe some modifications of leaves. (**23**)
14. What forms of asexual reproduction occur in flowering plants? (**25**)
15. What are the advantages of vegetative reproduction? (**26**)
16. Describe the structure of a flower. (**29–31**)
17. What is double fertilisation in flowering plants? (**33**)
18. What kinds of fruits are there? (**36**)
19. How may fruits and seeds be dispersed? (**37**)

20. What are the requirements for seed germination? (38)

21. Describe some of the kinds of storage organs in flowering plants. (40)

22. How may flowering plants be classified? (41)

CHAPTER VII

PLANT PHYSIOLOGY AND METABOLISM

1. Introduction. This chapter is concerned with how plants obtain and make food, how they utilise the energy contained in the food, how they grow, and how they respond to changing environmental conditions. In order to do this it is first necessary to summarise some features of the main chemical constituents of plants. This treatment will be brief, as more adequate discussion can be found in other HANDBOOKS (*see* Preface).

CHEMICAL CONSTITUENTS OF PLANTS

2. Water. One basic requirement for plants and other living organisms is water. There are a number of reasons for this.

(*a*) *Most chemical reactions which take place in living organisms occur in water.*

(*b*) *Water is a very good solvent* to the extent that very many chemicals dissolve in water, even if only to a limited extent. Even slightly soluble chemicals may be essential to plants, but may only be required in small quantities.

(*c*) *Water has a high heat capacity;* a large quantity of heat is required to raise a mass of water to a given temperature. Plants are even less able to moderate temperature changes internally than animals, and the heat capacity of water is useful to plants in enabling them to survive environmental extremes, especially as most chemical reactions within plants work well over a limited temperature range (*see* **13, 29** below).

(*d*) *Water has a maximum density at 4°C* so that ice floats and deep masses of water freeze from the surface downwards. While this is more important for aquatic animals than plants, the latter are less likely to freeze unless growing in the surface layers.

Water is thus the medium in which chemical reactions in living organisms take place, and the three main groups of organic (carbon-containing) chemicals involved in these reactions are:

(*a*) *carbohydrates;*
(*b*) *lipids* (fats);
(*c*) *proteins.*

3. Carbohydrates. The two main functions of plant carbohydrates are:

(*a*) *to act as an energy store*, e.g. starch;
(*b*) *to give structure*, e.g. cellulose in cell walls.

As well as starch and cellulose, carbohydrates include sugars of many kinds. Carbohydrates contain carbon (C), hydrogen

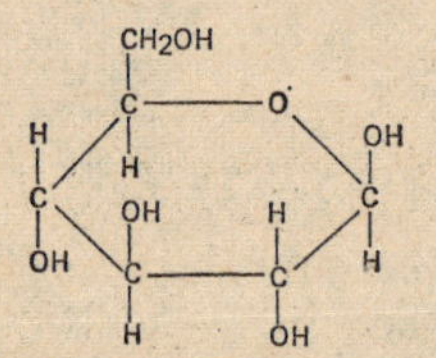

Fig. 44.—Ring structure of glucose molecule.

(H) and oxygen (O) and the ratio of hydrogen to oxygen is 2:1. With the chemical formula represented in the simplest way, a typical carbohydrate would be the sugar *glucose*, with the formula $C_6H_{12}O_6$. In practice, the carbon atoms usually form rings, so that a more true representation might be as seen in Fig. 44.

Carbohydrate molecules vary greatly in size and include some of the largest molecules known. Even so, they are normally based on a small number of simpler units or sugars, with a limited number of carbon atoms in each unit. Sugars include: *trioses* with three carbon atoms; *pentoses* such as ribose, with five carbon atoms; and *hexoses* like glucose, with six carbon atoms.

The basic sugar unit is termed a *saccharide* and there are different classes of saccharides.

(*a*) *Monosaccharides* include the examples given so far, all having one sugar unit.

(*b*) *Disaccharides* have two sugar units and include several important plant carbohydrates, including sucrose. A disaccharide may be formed from two monosaccharides, with a water molecule being lost in the process, so that the reaction is termed a *condensation reaction*:

$$2C_6H_{12}O_6 = C_{12}H_{22}O_{11} + H_2O$$

A disaccharide need not be composed of two six-carbon (hexose) sugars, but many important examples are, including *sucrose*, made up of *fructose* and glucose units, and *maltose*, made up of two glucose units. Both monosaccharides and disaccharides are soluble in water, form crystals, and are sweet to taste.

(*c*) *Polysaccharides* are larger carbohydrates and may have many hundreds of monosaccharide units of different kinds in their structure. Two of the most important plant polysaccharides are starch, a very common food storage molecule made up of glucose units, and cellulose, a key structural component of cell walls. Polysaccharides are not normally sweet, nor do they crystallise.

4. Uses of carbohydrates. Apart from important food storage and structural support functions, carbohydrates have other functions. Soluble sugars are easily transported within a plant, mainly in phloem tissues, and represent a means of transporting food from sites of synthesis to sites of storage and use. They also exert an *osmotic pressure* (*see* **18** below), especially important in roots in the uptake of water and mineral salts. The simpler carbohydrates such as glucose are some of the most readily available energy sources and are easily broken down or *oxidised* during *respiration* in order to provide energy for the work which a plant undertakes in living and reproducing.

5. The lipids are often known as *fats* and *oils*, the former solids and the latter liquids at room temperature. Lipids have a high energy content and have important roles as energy stores and also as structural components of living organisms. They contain carbon, hydrogen and oxygen, but proportionally much less oxygen than in carbohydrates and are composed of two sorts of chemicals, *fatty acids* and *glycerol*.

6. Chemical structure of lipids. Glycerol is an alcohol with the formula shown in Fig. 45. It combines with one, two or

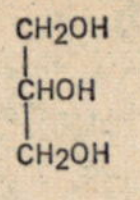

FIG. 45.—Glycerol formula.

three fatty acids in a condensation reaction to produce a fat or oil, depending on the kinds of fatty acids, the lipids being termed *mono-*, *di-* or *triglycerides* respectively. Thus one molecule of the fatty acid *oleic acid* can combine with glycerol to form a monoglyceride (see Fig. 46). Such a condensation

CH_2OH				$CH_2COOC_{17}H_{33}$		
|				|		
$CHOH$	+	$C_{17}H_{33}COOH$	→	$CHOH$	+	H_2O
|				|		
CH_2OH				CH_2OH		
glycerol		oleic acid		a monoglyceride		water

FIG. 46.—Condensation reaction to produce a monoglyceride.

reaction can take place in reverse, involving the splitting of the lipid into its constituents with the addition of water, the process being known as *hydrolysis*.

7. Functions of lipids.

(*a*) *Energy storage.* Because of their high energy content, lipids are common energy stores in seeds. Oil seeds such as soya bean, groundnut and the palm kernel are economic sources of vegetable lipids.

(*b*) *Structure.* The two main structural uses of lipids are in membranes within plant cells and as outer protective coats on plants, where they are particularly important in cutting down water loss. In the latter case, waxes are the most common lipids found. Lipids can combine with *phosphoric acid* to form *phospholipids* which have important specialised functions in membranes.

8. Proteins. The third important type of compound in plants is protein. Plant proteins also contain carbon, hydrogen and oxygen but can be far more complex in structure than either carbohydrates or lipids and always contain nitrogen (N) as well as frequently containing sulphur (S) and phosphorus (P).

9. Chemistry of proteins. The structural unit of the protein is the *amino acid* which has the general formula of

$$NH_2.R.COOH$$

where R can be a number of chemical units varying greatly in size, shape and structure. There are more than twenty naturally occurring amino acids and plants can usually make all they require, whereas most animals can only synthesise the simpler ones. Proteins consist of long chains of amino acids joined by chemical bonds termed *peptide linkages*. The many different amino acids in proteins can be arranged in chains in many different orders so that the number of different kinds of proteins is immense.

10. Functions of proteins.

(*a*) *Enzymes*. These are always partly, and sometimes entirely, made up of proteins. Enzymes are biological *catalysts* (*see* **13** below) which greatly speed up otherwise slow reactions without themselves being changed permanently.

(*b*) *Structure*. Proteins are important constituents of many plant structures. Membranes, for example, commonly include a protein component and this may have a control function, determining which molecules are allowed to pass through a membrane and even actively moving molecules across a membrane.

(*c*) *Pigments*. Plant pigments have essential roles in processes such as photosynthesis (*chlorophyll*) and respiration (*cytochromes*). Many plant pigments are partly protein.

(*d*) *Contraction*. Although movement is less important in plants than in animals, it is still vital in, for example, simple motile algae and motile gametes of many plants. Motility is dependent on contraction of organelles, and the contractile molecules are normally proteins.

11. Nucleic acids. There are many other important groups of chemical compounds found in plants, but probably most deserving of mention are the nucleic acids. These are large molecules with molecular weights of up to 100 000. Their main roles are in the storage of genetic information and its transmission within the cell to control the growth, development and activities of that cell. The function and replication of nucleic acids will be considered in Chapter VIII.

12. Structure of nucleic acids. Nucleic acids are composed of long chains of units called *nucleotides*, which consist of an organic base containing nitrogen linked to phosphoric acid and a pentose (5-carbon) sugar. The two kinds of acid are known as *deoxyribonucleic acid* or *deoxyribose nucleic acid* (DNA) and *ribonucleic acid* or *ribose nucleic acid* (RNA). The former contains the sugar deoxyribose and the four bases in its nucleotides are *cytosine*, *thymine*, *adenine* and *guanine*. RNA contains ribose as the sugar and the base thymine is replaced by another base called *uracil*. DNA is found in the nucleus of plant cells, is the store of genetic information and is able to replicate itself in cell division. RNA is concerned with transferring the genetic information from the nucleus to the rest of the cell and exists in a number of forms to perform this function including *messenger RNA*, *transfer RNA* and *ribosomal RNA*.

13. Enzymes. Although not strictly a class of chemicals, enzymes have such an important role in living organisms that they deserve attention at this stage. They normally consist mainly of protein but contain other compounds as well. In their role as biological catalysts they allow reactions to proceed at speeds which would otherwise be impossible at the low temperatures of living organisms. Without enzyme action biochemical reactions would be so slow that living organisms would not be able to respond to even slow changes in their environment. In addition to a protein component, enzymes may contain co-enzymes which can range from complex molecules such as *vitamins* of the *B* group to *heavy metals* such as copper, iron or molybdenum. There are several important properties of enzymes.

(*a*) *Turnover rate.* An enzyme molecule can work very rapidly, catalysing the change of *substrate molecules* into *product molecules* many times a second without itself being permanently changed. The turnover rate can be as high as 5 000 000 molecules per minute.

(*b*) *Specificity*. Enzymes will normally catalyse only one kind of reaction and may be specific for a single reaction.

(*c*) *Temperature response*. Enzymes operate best over the temperature range at which the cells in which they are found most commonly exist, and their efficiency decreases markedly outside this range. Enzymes rarely work well below 5°C or above 45°C but exceptions include enzymes in plants inhabiting hot springs with temperatures up to 65°C.

(*d*) *pH response*. Enzymes normally have an *optimum pH range* which, while usually close to neutrality, can for some enzymes be markedly acid or alkaline.

(*e*) *Reversibility*. Enzymes will catalyse a reaction in either direction. For example, reaction

$$A+B \rightleftharpoons C+D$$

may be catalysed by an enzyme in either direction until equilibrium is reached.

(*f*) *Enzyme inhibition*. Enzymes may be inhibited easily.

(*i*) *Competitive inhibitors* may not affect the enzyme molecule but compete with it in combining with substrate molecules, thereby preventing the enzyme from acting as a catalyst.

(*ii*) *Non-competitive inhibitors* prevent the enzyme molecules from acting permanently and include poisons such as arsenic and some heavy metals.

FORMS OF PLANT NUTRITION

14. The purpose of nutrition is to provide an organism with the chemicals it requires to survive, either by making them or by obtaining them from its external environment.

There are a number of forms of nutrition and plants use most of them.

(*a*) *Autotrophic nutrition*. In this, an organism makes its own food from simple inorganic components such as water and carbon dioxide, using an external energy source to make the food. Food thus produced effectively stores this external energy source.

(*i*) *Photosynthetic autotrophs* (photoautotrophs), such as green plants, use solar radiation as the energy source to make food by the process of photosynthesis. Almost all life on earth depends ultimately on this process.

(*ii*) *Chemosynthetic autotrophs* (chemoautotrophs) are relatively rare micro-organisms such as the sulphur bacteria which use chemical energy to make their own food.

(*b*) *Heterotrophic nutrition.* Here, organisms take in existing food produced by autotrophs. The most important heterotrophs are *holozoic heterotrophs*, those animals which feed on solid material, taking it in through a mouth or similar orifice. Plants, however, are included in the two other classes of heterotrophs.

(*i*) *Saprophytic heterotrophs* include most fungi and bacteria and a few higher plants. They take in dissolved material, dissolved often by enzymes which they secrete, the material usually being the decaying remains of other organisms.

(*ii*) *Parasitic heterotrophs* include parasitic fungi and bacteria which may attack many forms of plants and animals. Parasites feed directly on another organism and their nutrition may be extremely complex.

These nutritional boundaries are not completely defined. For example, many parasitic plants such as fungi can be saprophytic for part of their life history, assuming a parasitic style of nutrition if they chance upon a suitable host. Similarly some photosynthetic higher plants can act as parasites. In the concluding sections of this chapter, the major attention will be given to the nutrition of typical photosynthetic higher plants.

PLANT NUTRITION IN PHOTOSYNTHETIC AUTOTROPHS

15. Introduction. A typical photosynthetic autotroph requires light as an energy source and the following raw materials:

(*a*) *water;*
(*b*) *mineral salts;*
(*c*) *carbon dioxide.*

Water and mineral salts are obtained from the soil and carbon dioxide is taken in from the surrounding atmosphere.

16. Water is required:

(*a*) *as the medium for biochemical processes* within the plant;

(*b*) *as a transport medium*, most materials being transported within a plant in solution or aqueous suspension;

(*c*) *to provide chemical constituents* for many of the molecules which make up the components of the plant body;

(*d*) *for cooling the plant*, usually by evaporation from leaf surfaces.

The main site for the use of water is the leaf, for it is there that water is split into hydrogen and oxygen and the former used to reduce carbon dioxide to sugars via photosynthesis.

NOTE: Reduction, the opposite of oxidation, is effectively the removal of oxygen or the addition of hydrogen to a molecule, and usually requires an input of energy, part of that energy being stored by the reduced molecule and available for release by oxidation.

Plants, especially in hot weather, may require immense quantities of water; an oak forest, for example, uses at least 10 000 000 litres of water per hectare (ha) per annum.

17. Transpiration is the process by which plants obtain water from the soil for use, primarily in the leaves. The *transpiration stream* runs from the root hairs into the root, stem, petiole and leaf xylem elements and into the mesophyll cells of the leaf. Two forces are mainly involved:

(*a*) *root pressure;*
(*b*) *evaporation.*

Of the two, root pressure is by far the weaker force.

18. Root pressure pushes water up from the roots. It is primarily due to the effects of osmotic pressure in root hair and root cortex cells, pushing water into xylem elements.

NOTE: *Osmosis* takes place when molecules diffuse across a *semi-permeable membrane*. Larger molecules tend to stay on one side of the membrane whereas smaller molecules such as water can pass through and dilute a strong solution of larger molecules. Membranes of root hair cells act as semi-permeable membranes between water in the soil and the contents of the cells. The cells have a high concentration of large sugar molecules and tend to get diluted by soil water. The water accumulates in the roots and is pushed through into the xylem vessels.

19. Evaporation. This provides the main force by which water moves up a plant into the leaves. It is a result of diffusion of water molecules from the saturated atmosphere within a leaf through the stomata and out into the surrounding atmosphere. Even in only moderately dry atmospheres, transpiration is a powerful force and for most land plants the problem is one of losing too much water rather than difficulties in bringing it up from the roots. In fact an important aspect of the colonisation of dry land was the ability of plants to develop large photosynthetic areas which did not desiccate quickly.

NOTE: Forces of transpiration can exceed 1 atmosphere without a vacuum forming in the vessels. This is because the columns of water in xylem vessels are only 0·5 mm in diameter. Such columns have considerable tensile strength and are not easily broken. Transpiration can therefore "drag" water up to the top of a tall tree.

Water loss from leaves is increased by:

(*a*) *high temperature;*
(*b*) *low relative humidity;*
(*c*) *wind.*

In addition to the normal waxy covering of the cuticle and the control of stomatal opening, many plants have further modifications to restrict water loss. These include:

(*a*) *stomata sunk in pits and few in number;*
(*b*) *fleshy leaves* with a large volume relative to surface area;
(*c*) *hairy leaves* slowing down movement of air;
(*d*) *a thick cuticle.*

20. Wilting. There is normally a continuous column of water between roots and leaves, stationary at night when stomata are closed and moving during the day. If the rate of transpiration exceeds the rate of movement of water uptake from the soil, or the rate at which it can be moved through the xylem, leaf mesophyll cells lose water, lose their turgor and collapse (with the cell membrane shrinking inwards and no longer giving support to the thin cell wall). The leaf then wilts. Temporary wilting is common in plants but beyond the *permanent wilting point* there is irreversible damage and plant cells die.

NOTE: In cool damp conditions where the *atmospheric relative humidity* equals that of the interior of the leaf, evaporation and therefore transpiration ceases, but without harming the plant.

21. Mineral salts. A wide range of minerals is required by plants, three of them, nitrogen, phosphorus and potassium, being required in quite large amounts. All the minerals are normally obtained by the uptake of mineral salts from the soil.

NOTE: *Legumes* and a few other plants have symbiotic bacteria present in *root nodules* which can "fix" atmospheric nitrogen into nitrogen-containing organic compounds which are available to the plant.

Within nature, the main mineral nutrients are cycled by processes termed *biogeochemical cycles* (*see* IX, 12) and these help to ensure the continued availability of mineral elements to plants.

22. Uptake of minerals takes place mainly through root hairs and involves two processes:

(*a*) *Passive transport*, where the *mineral ions* diffuse freely into the roots, usually as part of the uptake of water;
(*b*) *Active transport* where *ion pumps* in the membranes of root hair and other root cells pump mineral ions across the membranes and into the cells, a process which requires an energy input.

Mineral nutrients are grouped into macronutrients and micronutrients, or trace elements, the latter being required in very small amounts. Table I lists the more important mineral nutrients with their functions and deficiency symptoms.

NOTE: Plants also require zinc (Zn), copper (Cu), cobalt (Co) and molybdenum (Mo) in very small amounts and some plants require chlorine (Cl), sodium (Na), silicon (Si) and iodine (I).

23. Occurrence of deficiency. Of all the mineral elements mentioned, nitrogen, phosphorus and potassium are most often limiting to plant growth and these are the constituents of artificial fertilisers. Deficiency of other mineral nutrients is rare, except for a deficiency of iron, copper and manganese, which can be caused by an alkaline soil, these mineral nutrients being rather insoluble under alkaline conditions. This is often

TABLE I: THE MORE IMPORTANT PLANT MINERAL NUTRIENTS

Element	*Function*	*Deficiency symptoms*
MACRONUTRIENTS		
Nitrogen (N)	Constituent of proteins and other compounds	Chlorosis (leaf yellowing) and poor growth
Potassium (K)	Activates enzymes and is used in enzyme synthesis	Chlorosis
Calcium (Ca)	Constituent of membranes and middle lamella	Deformed and poor growth
Phosphorus (P)	In nucleic acids, ATP and many other compounds	Poor transport, respiration and growth
Sulphur (S)	In certain proteins and lipid metabolism	Chlorosis and poor shoot growth
Magnesium (Mg)	Constituent of enzymes and chlorophyll	Chlorosis and reduced growth
MICRONUTRIENTS		
Iron (Fe)	Constituent of enzymes in respiration and chlorophyll synthesis	Chlorosis and poor growth
Boron (B)	Probably essential for cell walls and sugar trans-location	Growth distortions
Manganese (Mn)	Constituent of many enzymes	Chlorosis and poor growth

known as a *lime-induced chlorosis* and is most frequent in roses and apples.

24. Carbon dioxide is the most readily available raw material required by plants and is present in the atmosphere at a concentration of about 320 parts per million (p.p.m.). Its concentration varies at different times of the day and at diferent seasons of the year and the atmospheric concentration is ising slightly each year through the burning of fossil fuels. None of these variations appears to be sufficient to affect plant growth.

Carbon dioxide enters the plant via stomata and diffuses through the spongy mesophyll cells, going into solution at the surface of these and the palisade mesophyll cells.

25. Photosynthesis has been described as the most important chemical process in the world; it utilises solar radiation to provide food for the plant and animal kingdoms. The process may be summarised:

$$6CO_2 + 6H_2O \xrightarrow{\text{Light Energy}} C_3H_{12}O_3 + 6O_2$$

but while this shows the initial and final products, it is misleading in not indicating the complexity of the processes involved.

26. The light and dark reactions. There are actually two groups of linked reactions in photosynthesis:

(*a*) *The light reaction* which actually involves trapping light energy and therefore takes place in the light;

(*b*) *The dark reaction* in which carbon dioxide is fixed to form organic compounds. It is called the dark reaction because it does not require light, but can take place in the light or in the dark.

27. The light reaction. Light energy is initially trapped by many different plant pigments in the grana of chloroplasts in photosynthetic cells. The pigments include carotene and different kinds of chlorophyll but all pass on the energy to a particular kind of chlorophyll which is said to become "excited". In this excited state it is capable of passing on its energy by releasing high energy *electrons* which are then used for two purposes:

(*a*) *To break down water into hydrogen and oxygen*, the former being taken up by an organic molecule called *nicotinamide adenine dinucleotide phosphate* (NADP) to form reduced nicotinamide adenine dinucleotide phosphate ($NADPH_2$) and the latter being given off as molecular oxygen into the atmosphere;

(*b*) *To form an energy-rich compound*, called *adenosine triphosphate* (ATP) from adenosine diphosphate (ADP) and inorganic phosphate (P).

NOTE: ATP consists of the base adenine, connected to a molecule of the sugar ribose and three inorganic phosphate groups. It is very important to living organisms because of its high energy level and is used as a basic unit of "energy currency".

28. The dark reaction. ATP and $NADPH_2$ both provide the energy for the dark reaction. It is basically a cyclic process called the *Calvin cycle* after the American chemist Melvin Calvin who did much of the work on it in the 1940s and 1950s.

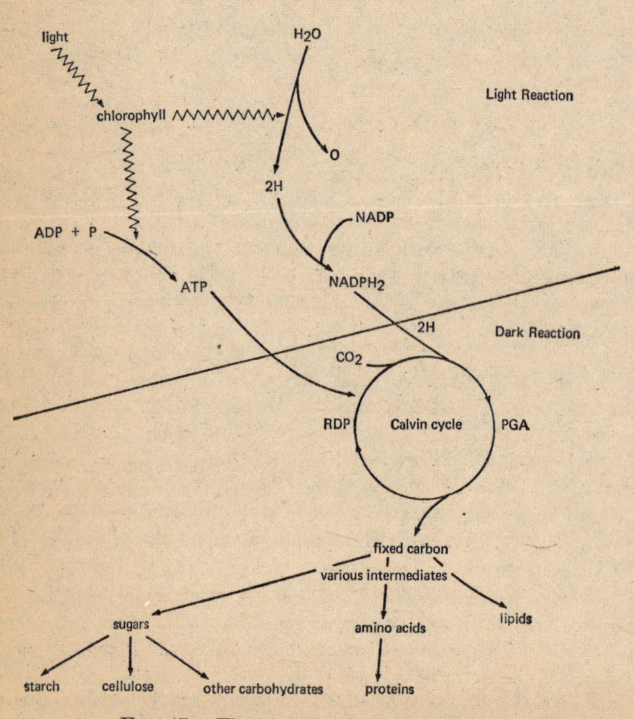

FIG. 47.—The process of photosynthesis.

As shown in Fig. 47, a 5-carbon sugar derivative called *ribulose diphosphate* (RDP) joins with a molecule of carbon dioxide to form two molecules of the 3-carbon compound *phosphoglyceric acid* using the energy of ATP. The phosphoglyceric acid is then reduced to *phosphoglyceraldehyde* (PGA) by hydrogen atoms from $NADPH_2$. This fixing of carbon requires a lot of energy and is termed *endergonic* (or *endothermic*).

From the PGA, various other carbon compounds are produced, many of which serve as intermediates in the synthesis of carbohydrates, proteins and lipids. However, some of the PGA is used to make new RDP so that more carbon dioxide can be fixed and the cycle keeps on turning. The light reaction and the Calvin cycle take place in the chloroplast.

NOTE: This is a brief summary. All the processes described involve intermediates and there are scores of different compounds involved in photosynthesis.

In the summary equation usually given for photosynthesis (*see* 25 above) a 6-carbon sugar such as glucose was the product. This is a reasonable summary as much of the PGA is actually used to make such sugars.

29. Factors affecting photosynthesis. All the inputs into the photosynthetic process such as light energy, carbon dioxide and water will obviously affect the rate of photosynthesis if they are not present in the right concentrations. Shade-tolerant plants (*see* IX, 8) need less light than ordinary plants, but most plants can usefully use much more than the 0·032 per cent concentration of carbon dioxide available to them in the atmosphere. For this reason, carbon dioxide enrichment of glasshouses is practised in commercial horticulture for increasing the yields of glasshouse crops.

Regarding temperature, the enzymes involved in catalysing parts of the dark reaction work best over the range 10°C to 40°C, with an optimum near 40°C so that within that range, the higher the temperature, the higher the rate of photosynthesis.

TRANSLOCATION

30. The aim of translocation. Photosynthesis fixes carbon to form a variety of compounds essential to the plant and these must be moved to those places within the plant where they are required either for growth, storage or provision of energy. This is the process of *translocation* which takes place through phloem tissues (*see* VI, 9) which permeate all the major parts of a plant.

At different times in the life of a plant, products of photosynthesis move from their *source* (photosynthetic tissue) to *sinks* where they are used. Sinks include apical and secondary meristems and associated areas of cell enlargement in the stem and root, enlarging storage organs and especially flowers, seeds and fruits.

31. The mechanism of translocation is not clearly understood but is rarely, if ever, by passive diffusion as it normally takes place much too fast for that. Phloem, unlike xylem, is a living tissue and active transport processes are probably involved.

RESPIRATION

32. Respiration in plants. Plants, like all living organisms, require a continuous source of energy (i.e. the power to do work) for growth, development and reproduction. The energy is acquired through *respiration*, the breakdown of sugars into carbon dioxide and water with the release of energy. It is

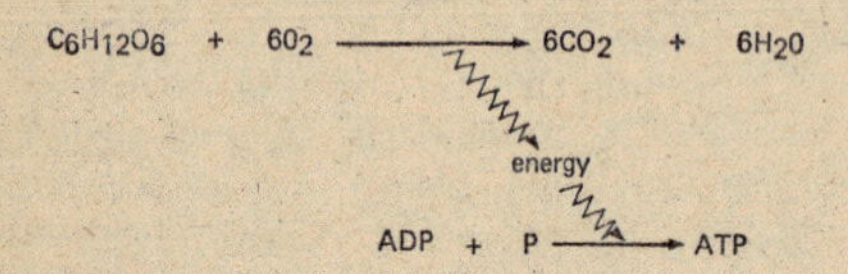

FIG. 48.—A representation of the process of respiration.

therefore the reverse of photosynthesis and is an *exergonic* (or *exothermic*) process and much of the energy released is conserved by the formation of ATP, which is then available to power many other processes. Respiration can therefore be represented as in Fig. 48. About half of the energy released is actually conserved in the formation of energy-rich ATP, the remainder being wasted.

33. Anaerobic respiration. Although efficient, respiration has a disadvantage in that it requires oxygen. In practice, the irst part of respiration, *glycolysis* or the partial breakdown of lucose, does not require oxygen, and some organisms only ise this process and produce carbon dioxide and a waste roduct, such as ethyl alcohol:

$$C_6H_{12}O_6 \rightarrow 2C_2H_5OH + 2CO_2$$

This is termed anaerobic respiration and allows organisms which use it to live in environments deficient in oxygen. Unfortunately it is very inefficient as only 3 per cent of the nergy is conserved as ATP. Fortunately, the waste product is lcohol which can be put to good use by higher animals!

34. Aerobic respiration. In the full process of respiration, glycolysis results in pyruvic acid rather than ethyl alcohol and this is fed into the second part of the process known as the *Krebs cycle* after Sir Hans Krebs who discovered it over 30 years ago. It may be summarised as in Fig. 49. This is a

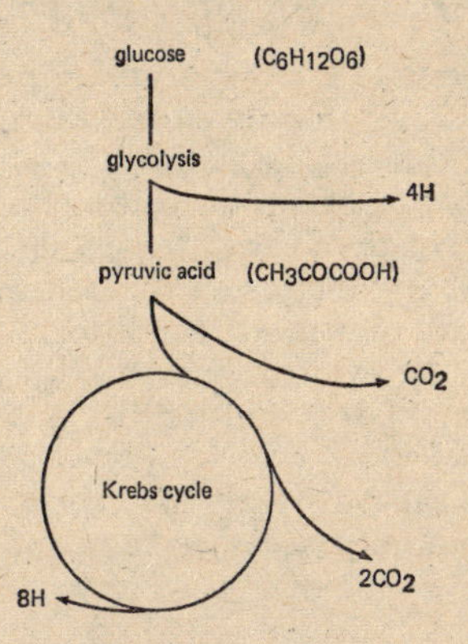

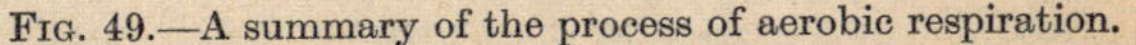
FIG. 49.—A summary of the process of aerobic respiration.

simplification as the Krebs cycle and glycolysis both involve many intermediate stages. While some ATP is formed directly during the process, most is formed when hydrogen atoms, given off in glycolysis and in the Krebs cycle, pass down a *respiratory chain* of enzymes. At the end of this they unite with oxygen to form water, but in the process they give off energy which is conserved in ATP, formed from ADP and P.

35. Site of respiration. Glycolysis takes place in the cell cytoplasm but the Krebs cycle is restricted to the mitochondria with the respiratory chain enzymes located on the inner membrane of the mitochondria (*see* I, **10**).

36. The compensation point. A plant produces food by photosynthesis and consumes it in respiration. The *compensation point* is reached when the rate of production equals the rate of consumption, usually at a low light intensity. During the day, production normally exceeds consumption and during the night the reverse is true.

NOTE: Plants obtain oxygen by free diffusion from the atmosphere and a limited amount of diffusion of dissolved oxygen within the plant. However, roots in waterlogged soil may not get sufficient oxygen in this way and can experience oxygen starvation which, in extreme cases, will kill the roots.

CONTROL OF GROWTH

37. Plant hormones. Photosynthesis provides the food for green plants but they will only succeed in growing and reproducing if these processes are controlled in some manner. Roots growing into the air, shoots growing downwards, flowers formed at the wrong time of the year or leaves shed in the spring are all usually counter-productive. Given adequate food supplies and tolerable environmental conditions, the growth and development of a plant is largely controlled by *hormones*, often termed *phytohormones*, to avoid confusion with those produced by animals.

Phytohormones act at very low concentrations and are usually produced in one part of a plant but affect tissues elsewhere. There are several groups of phytohormones.

38. Auxins affect cell growth rather than division and an important auxin is *indoleacetic acid* (IAA). The auxins help to control secondary growth such as lateral root formation, wound healing, abscission and inhibition of buds as well as the *tropisms* of a plant.

NOTE: A tropism is a response to a stimulus such as gravity or light and can be positive or negative. e.g. a root might be *positively geotropic*, growing downwards in response to gravity, and also *negatively phototrophic*, growing away from light.

Auxins tend to be synthesised in apical meristems and affect elongating tissues near these meristems.

39. Gibberellins are probably best known for their ability to produce very tall cabbages but they are now believed to be very important in the general control of a plant's growth. For example, gibberellins affect the relative rate of growth of stem and leaf and help determine whether a plant will be tall and thin or short and compact, perhaps in response to different wind conditions. There are over thirty closely related gibberellins.

40. Phytochromes are hormones which respond mainly to light. Their main function is to control the time of flowering and they will respond to changes in day-length, producing a flower-inducing hormone called *florigen* in order to ensure the production of flowers at the best time of year. Phytochromes can also affect seed germination and stem elongation.

41. Kinins include *kinetin* and other hormones and are primarily concerned with controlling the rates of cell division.

42. Ethylene, a gas, is probably the simplest plant hormone (C_2H_4) and tends to act largely by inhibiting various processes, including the action of some other hormones such as auxins. It will stimulate fruit ripening and will act at concentrations in air as low as 1 part per ten million.

43. Hormone interaction. Relatively few phytohormones control an immense amount of plant activity. This is partly accomplished by the interaction of hormones. For example, auxins interact with kinins and ethylene with auxins. This factor of interaction allows for complex controlling activities by plant hormones.

PROGRESS TEST 7

1. Why is water important to plants? (2)
2. What are the functions of carbohydrates in plants? (3)
3. What are polysaccharides? (3)
4. What role do lipids play in plant structure? (7)
5. What are the functions of proteins in plants? (10)
6. What important properties do enzymes have? (13)

7. Describe the different kinds of plant nutrition. (14)
8. What raw materials does a green plant require? (15)
9. What is root pressure? (18)
10. What means do plants have for avoiding water loss? (19
11. What is the permanent wilting point? (20)
12. What are the functions of the macronutrients? (22)
13. What is lime-induced deficiency? (23)
14. Why is the dark reaction so called? (26)
15. How does temperature affect the rate of photosynthesis? (29)
16. What is translocation? (30)
17. Why do plants respire? (32)
18. What is glycolysis? (33)
19. What is the function of the respiratory chain? (34)
20. Where does the Krebs Cycle take place in the plant cell? (35)
21. What is the compensation point? (36)
22. Name an auxin. (38)
23. What is a tropism? (38)
24. Why is hormone interaction important? (43)

CHAPTER VIII

GENETICS AND EVOLUTION

INHERITANCE AND GENETICS

1. Inheritance in plants was first investigated by Gregor Mendel, an Augustinian monk living in Austria (1882–84). He studied the way in which characters of the garden pea plant were transmitted from one generation to the next. He was one of the founders of *genetics*, the study of inheritance.

2. The monohybrid cross. One of the characters Mendel investigated was the height of the pea plants. If a tall plant is self-pollinated and all the seeds produce tall plants, then this seed is said to be of a *pure-breeding* tall line. Similarly, if dwarf plants, when self-pollinated, produce seed which gives all dwarf plants, then the seed is of a pure-breeding dwarf line.

When however plants of these lines are crossed together, no matter which way this is done, with tall or dwarf plants producing the pollen or the ovules, then all the seed produced develops into tall plants. This is called the *first filial generation* or F_1 hybrids. When, however, these F_1 tall plants are self-pollinated, the seed produces the *second filial generation* or F_2 generation which has both tall and dwarf plants in the ratio of 3 tall to 1 dwarf.

The explanation for this is that pea plants are diploid and have two sets of chromosomes and both chromosomes of a particular (*homologous*) pair carry a factor called a *gene* for height. In pure-breeding tall plants, both chromosomes carry genes for tallness and are said to be *homozygous* for tallness. This may be represented by the notation TT (*see* note below). Similarly in a pure-breeding dwarf plant, both chromosomes of the particular (homologous) pair carry genes for dwarfness and are homozygous for dwarfness, represented by tt.

The haploid gametes, pollen grains or ovules, however, carry only one chromosome of a particular pair and therefore have only one gene. Gametes from the tall plants are all alike and have one gene for tallness T. Gametes from the dwarf plants are also all alike and have one gene for dwarfness t.

When these gametes fuse to form a new individual, the F_1 generation, there are again two chromosomes each with a gene for height, one from the tall plant T and one from the dwarf plant t. The resulting plant, represented by Tt, is said to be *heterozygous* for tallness and is tall. It has therefore adopted the character of only *one* of the genes which is therefore said to

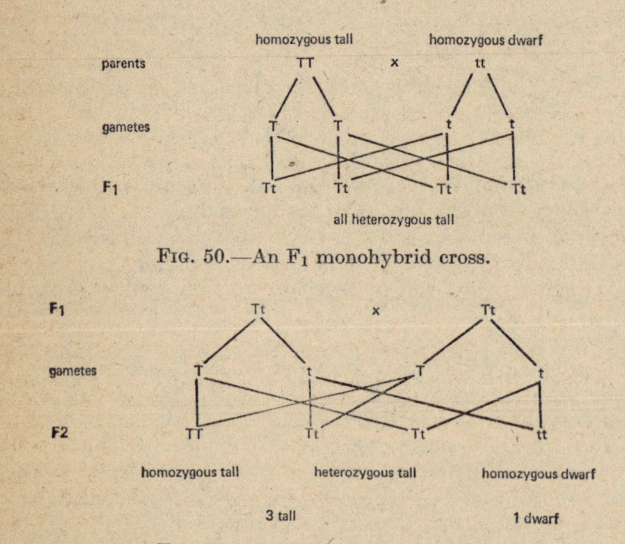

FIG. 50.—An F_1 monohybrid cross.

FIG. 51.—An F_2 monohybrid cross.

be *dominant*. The other, which does not manifest itself in the appearance of the plant is said to be *recessive*.

When these F_1 hybrids Tt are self-pollinated they produce gametes of two types T and t, so that at fertilisation a gamete with a gene for tallness T may fuse with a similar gamete T to give TT, a pure breeding tall plant, or it might fuse with a gamete carrying a dwarf gene t to give Tt, a tall but heterozygous plant and not pure-breeding.

The other gamete carrying the dwarf gene t may fuse with a gamete carrying a tall gene T, again to produce a heterozygous tall plant represented by Tt. It may, however, fuse with a gamete carrying a gene for dwarfness t. This develops into a dwarf plant which is homozygous for dwarfness tt.

This is how the ratio of 3 tall to 1 dwarf is achieved in the F_2 generation; of the three tall plants, only one is homozygous and the other two are heterozygous.

The F_1 cross is represented diagrammatically in Fig. 50, and the F_2 cross is therefore represented diagrammatically as in Fig. 51.

NOTE: In genetic notation, a capital letter represents a dominant gene and the small letter represents a recessive gene. In heterozygous material, the dominant precedes the recessive (i.e. Tt). In a dihybrid cross (*see* 4 below) letters representing different genes or characters are separated.

3. The back-cross. There is no way the genetical composition of a plant can be told from its appearance, i.e. a tall homozygous plant looks just like a tall heterozygous plant. The way in

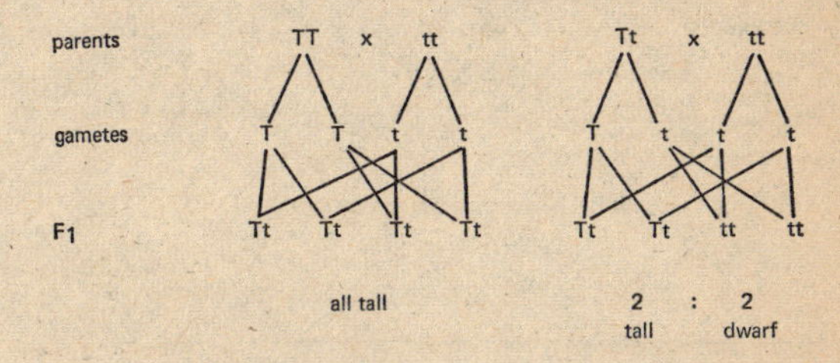

FIG. 52.—A back-cross to determine genetic composition.

which the genetical composition can be determined is by crossing the plants with a homozygous recessive dwarf plant. This is called a *back-cross*.

Homozygous tall plants crossed with homozygous dwarf plants result in all the F_1 generation being tall and heterozygous. Heterozygous tall plants crossed with homozygous dwarf plants result in the F_1 plants being half tall and heterozygous and half dwarf (*see* Fig. 52).

4. The dihybrid cross. It is possible, provided the characters which are to be studied are sited on different chromosomes, to study two characters at the same time. Again the classic

experiments were conducted by Gregor Mendel on the garde
pea. He noticed that some plants produced round seeds whil
others produced wrinkled seeds, and both these seeds, th
round and the wrinkled, could be yellow or green in colour.

He selected a pure breeding round yellow seeded plant an
crossed it with a pure breeding wrinkled green seeded plant
All the F_1 plants were round and yellow seeded so that roun
and yellow were the dominant characters over the recessiv

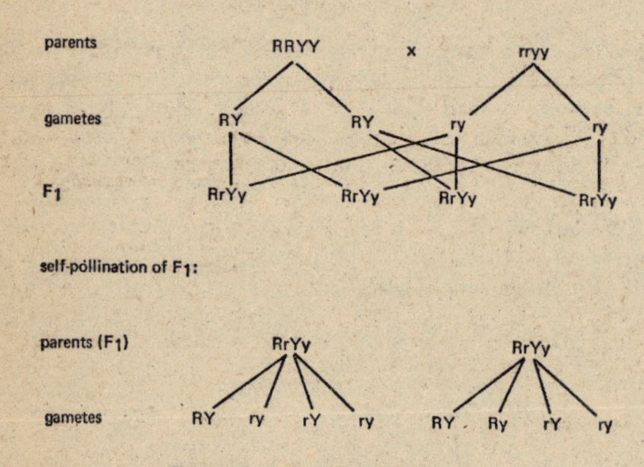

FIG. 53.—A dihybrid cross.

TABLE II: THE DIHYBRID CROSS

	RY	Ry	rY	ry
RY	**RRYY** Round Yellow	**RRYy** Round Yellow	**RrYY** Round Yellow	**RrYy** Round Yellow
Ry	**RRyY** Round Yellow	**RRyy** Round Green	**RrYy** Round Yellow	**Rryy** Round Green
rY	**RrYY** Round Yellow	**RrYy** Round Yellow	**rrYY** Wrinkled Yellow	**rrYy** Wrinkled Yellow
ry	**RrYy** Round Yellow	**Rryy** Round Green	**rrYy** Wrinkled Yellow	**rryy** Wrinkled Green

wrinkled and green characters. Therefore the parents were RRYY round and yellow and rryy wrinkled and green. The F_1 generation was therefore RrYy.

When the individuals of the F_1 generation were self-pollinated, the offspring appeared in the proportions of

9 round and yellow seeded
3 round and green seeded
3 wrinkled and yellow seeded
1 wrinkled and green seeded

as shown diagrammatically in Fig. 53.

It is easier now to display all the possibilities (*see* Table II).

5. Economic importance of plant genetics. The science of genetics is the basis for plant breeding, this being the incorporation of desirable characters into one particular plant stock. Desirable characters may include disease and drought resistance, high yield, special flower colour or even the most suitable shape of fruit for marketing, picking or canning.

The principles of plant breeding are fairly simple but there are many practical difficulties which make it less straightforward. The first thing the plant breeder has to do is to find a plant or plants which already have the desirable character, even if all their other characters are entirely undesirable. This is why plant breeding stations have large "museums of varieties" and send out expeditions to bring back wild and seemingly inferior types of crop species.

The wild plants with desirable characters are then crossed with crop plants and the offspring or progeny are screened to select improved varieties and these are then tested by back-crosses to ensure pure breeding lines. After this, the seed of the new variety is multiplied up to produce enough for passing on to the growers. The whole process can take many hundreds or even thousands of crosses and many years of testing and multiplication before the seed is available on the market.

THE NATURE OF THE GENE

6. DNA. The genetic material of plants in the DNA found in the nucleus (*see* I, **10**). The DNA molecule is made up of molecules of phosphate and sugar and four bases, adenine (A), thymine (T), cytosine (C) and guanine (G). These molecules are arranged in a ladder-like structure, the sides of which are twisted around each other to form a *double helix*. The sides of the ladder are made up of phosphate and sugar molecules whilst the rungs consist of complementary pairs of bases. The base pairs are held together by weak *hydrogen bonds*. Adenine and thymine constitute one pair whilst cytosine and guanine make up the other pair. Adenine always pairs with thymine and cytosine always pairs with guanine due to the configuration or shape of the molecules (*see* Fig. 54).

The order in which the pairs of bases occur along the length of the DNA molecule provides the code for the production of

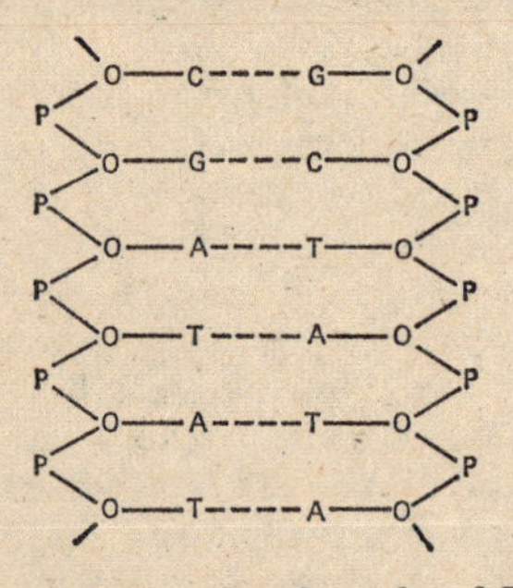

FIG. 54.—Diagram representing the formula of DNA. A = adenine; T = thymine; C = cytosine; G = guanine; P = phosphate; O = sugar.

different enzymes which can, in turn, control the cell's activities. The DNA molecules function in units of three bases called *triplets*. Each triplet conveys information to the cell and particular triplets, called *codons*, denote amino acids, the constituents of proteins. The particular codons for all amino acids have now been worked out, e.g. CGA codes for the amino acid alanine and TTA codes for asparagine. By containing this information, DNA is therefore the main material conveying the information for inheritance. It is normally found in the nuclei of cells, where it is formed into chromosomes. Chromosomes consist of two chromatids which are each made up of very long coiled and recoiled molecules of DNA.

7. Replication of DNA. It is thought that this is brought about by the uncoiling of the DNA molecules and the breaking of the weak hydrogen bonds holding the bases together. The two halves of the DNA molecule separate and complementary bases which are readily available within the cell latch on to the free bases of the separated helixes. The rest of the structure is then made up of sugar and phosphate molecules. Two identical molecules are formed because each half of the original molecule is used to form the template upon which to build the other half (*see* Fig. 55).

This replication of DNA takes place during the interphase between mitotic divisions. When the chromosomes become visible again, at the beginning of the next prophase, it is found that each chromatid is made up of two strands of DNA. At metaphase, the double chromatids separate and become the

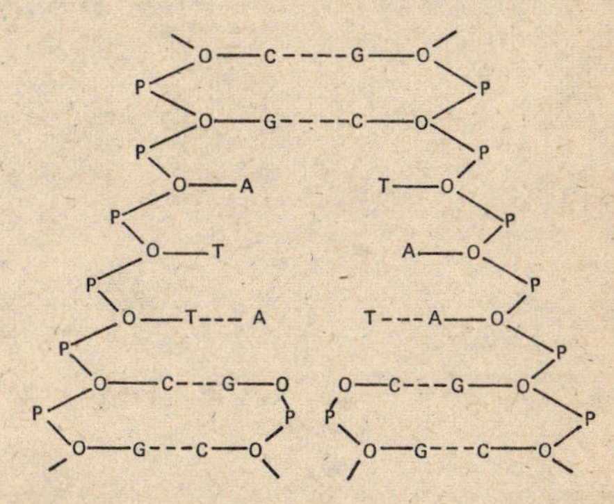

Fig. 55.—Diagram illustrating replication of DNA molecule.

chromosomes of the new cells. Each of the new cells therefore has the same genetic information as the parent cell.

8. Protein synthesis. A certain number of triplets along the length of the DNA molecule constitute a *gene*. As each triplet codes for an amino acid, genes code for whole proteins. The most important protein-based substances in the cell are the enzymes as they are fundamental to cellular activity. Genes can, therefore, code for specific enzymes.

Protein production takes place on the ribosomes of the rough endoplasmic reticulum. Therefore there has to be a method of carrying the information from the DNA in the nucleus to the ribosomes in the cytoplasm.

9. Messenger RNA. This is brought about by another nucleic acid called *messenger RNA*. RNA contains the bases adenine uracil, cytosine and guanine, the base thymine found in DNA

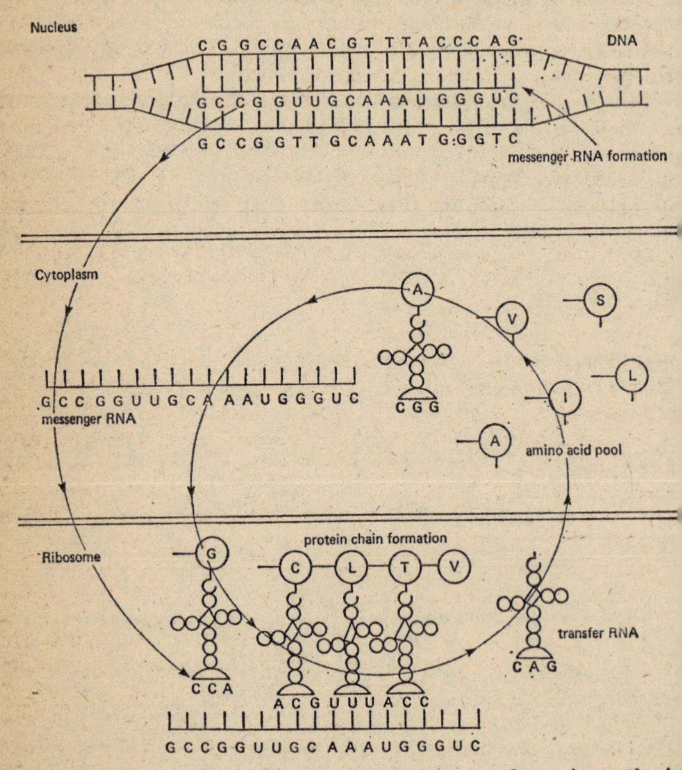

FIG. 56.—Diagram illustrating the general features of protein synthesis

being replaced by uracil. Ribose replaces deoxyribose as the sugar.

In the nucleus, the DNA double helix breaks and unwinds over the distance of the gene. Then, using only one of the two unwound strands, RNA bases and sugars are latched on from a pool in the cell. A length of messenger RNA is then formed on the DNA gene. This messenger RNA is released and the DNA re-forms and returns to its original state to be used again. The information has therefore been transferred from the DNA to the RNA.

The messenger RNA then passes out of the nucleus and makes its way to the ribosome, where it slides into a groove on the surface of the ribosome. Ribosomes are able to move along the messenger RNA as protein synthesis proceeds.

10. Transfer RNA. The amino acids which provide the basic units of proteins are brought to the ribosomes by another type of RNA called *transfer RNA*. This is made up of a single strand of RNA which is coiled up to form a "clover leaf" shaped molecule. The end of the "stalk of the leaf" is the point of attachment for an amino acid and on the end of the "middle leaflet" are three unpaired bases forming an *anticodon* which are specific for each amino acid.

The transfer RNA picks up a free amino acid in the cytoplasm and travels to the ribosome. The anticodon then fits into complementary bases on the messenger RNA and the amino acid is held in position at the other end of the molecule. A further transfer RNA molecule, bringing another amino acid, arrives at the ribosome and fits into the next codon. The two amino acids then become linked together. Once the amino acids are joined together to form a protein the transfer RNA is released and can be reused by the cell (*see* Fig. 56).

EVOLUTION

11. The nature of evolution. Evolution is the process by which changes in a population are perpetuated and may result in the formation of new species. The present day theory of how evolution occurs is based on the work of Charles Darwin (1809–82). Darwin stated that new species were developed by the natural selection of certain different individuals within an existing population. Previously it had been thought that each species was the result of a special creation.

Evolution is a continuous process but the rate at which it takes place varies depending upon the availability of the variation and the changes which occur in the environment. For the evolution of a new species to occur, there should be:

(*a*) *variation within the population;*
(*b*) *a change in the environment;*

(*c*) *competition;*
(*d*) *selection;*
(*e*) *formation of species.*

12. Variation. The variation within a population or *deme* is limited by the genes found within that population. This is called the *gene pool.* Not all the genes manifest themselves in the individuals of the population at once as some of the genes are *recessive.* These, however, do become apparent at some time if the plants are reproduced sexually, as crossing-over and fertilisation bring about a shuffling of the characters and the production of new combinations. The frequency with which a particular type of gene occurs compared with all the other types of genes for the same character is known as *gene frequency.*

13. Mutations. Further variation may be brought about by *mutations.* These are changes in the genetic material of the cells. Mutations occur naturally and for any one type of gene it has been calculated that one gene in 500 000 will mutate in a particular generation. There are two types of mutation.

(*a*) *Somatic mutations.* Mutations which involve the genetic material of the cells of the body of the plant are called *somatic mutations.* No matter how advantageous these mutations are to the individual which carries them, they are of no evolutionary significance unless the plant can be vegetatively propagated.

(*b*) *Other mutations* which affect the reproductive cells are easily passed on in the seed and may be of great evolutionary significance.

Mutations may involve single genes, groups of genes, whole chromosomes or complete sets of chromosomes. Most mutations involving single genes, groups of genes or whole chromosomes are deleterious to the particular population in which they occur. If, however, there is a change in the circumstances of the population, then they may be of advantage.

On the whole, single gene mutations are less deleterious than mutations of groups of genes or whole chromosomes. Occasionally, however, single gene mutations have very serious effects, causing the death of the plant. These are called *lethal genes.*

Whole sets of chromosomes may be duplicated and plants with more than two sets of chromosomes are called *polyploids* and whilst most mutations are not beneficial, polyploidy may produce larger plants with increased vigour and in some cases it can make infertile plants fertile. Many crop plants are polyploids including wheat, bananas, coffee, apples, peanuts and sugar cane.

14. Change in the environment. Changes in the environment fall into three groups.

(*a*) *Climatic changes* may be changes in rainfall, snow, temperature, amount of sunlight or exposure to wind. Atmospheric changes likely to affect plants would be changes in the content of oxygen or carbon dioxide of the atmosphere.

(*b*) *Edaphic factors* are those concerned with the soil and may include changes in soil water, oxygen and nutrient availability.

(*c*) *Biotic factors* may include changes in the level of predation by animals and human beings which might be experienced by the population. Disease is a biotic factor and perhaps one of the most important pressures for change.

15. Competition. Once the environment changes there develops within the population a struggle to take advantage of the new situation. The variation which is present within the population then becomes very important as some plants will be able to compete more effectively than others.

Occasionally it is not necessary to have a change in the environment for an increase in competition to take place, as some spontaneous mutations are advantageous in a given environment and the plants with the mutation will be more successful in that existing environment.

16. Selection. The result of competition is that the better adapted plants will reach maturity quicker and will reproduce more effectively than the rest of the population. Over a few generations, these plants will increase their proportion of the population whilst the losers will decline in numbers or may be forced into a different habitat. This is what Darwin called the "survival of the fittest" ("fittest" being the best fitted for the situation rather than the most physically agile).

17. Formation of species. A species is a group of similar individuals which will only breed within the group. Therefore, to form a new species there has to be a split between one group and another which stops them interbreeding. This may be brought about by mutations which stop cross-pollination, perhaps by altering the time of pollen production relative to the original population or by making plants self-fertile and unable to accept pollen from the original population. An alternative might be a lack of compatibility of the pollen and the abortion of the embryos. All these possibilities can lead to species formation.

Plants may also become physically separated and then different mutations in each group will eventually result in the production of two groups which will not interbreed.

Eventually the new species stabilises in its new environment, the gene frequency is found to be stable and the population is said to be in *genetic equilibrium* until the next change, whether it be inheritable or a change in the environment.

MAJOR TRENDS IN EVOLUTION

18. The major evolutionary trends that have taken place in the plant kingdom over geological time have been movement from:

(*a*) single celled plants to multicellular plants;
(*b*) self motile plants to stationary, fixed plants;
(*c*) aquatic plants to terrestrial plants;
(*d*) dependence of plants on water for fertilisation to plants independent of water;
(*e*) simple unorganised plants to those showing differentiation of tissues to give division of labour;
(*f*) the alternation of a large independent gametophyte with a small independent sporophyte plant to the alternation of a small gametophyte with a large sporophyte on which it is dependent.

It is now possible for evolution to be controlled by plant breeders in their search for new types of plant by inducing variation by *mutagens* (mutation inducing agents) and by careful selection of the progeny.

PROGRESS TEST 8

1. What is a monohybrid cross? (2)
2. Explain the terms homozygous and heterozygous. (2)
3. What is a back-cross? (3)
4. What is the composition of the DNA molecule? (6)
5. What is messenger RNA? (9)
6. How does transfer RNA work? (10)
7. What is the significance of variation in evolution? (12)
8. What is a mutation? (13)
9. What are the major evolutionary trends in the plant kingdom? (18)

CHAPTER IX

PLANT ECOLOGY

THE NATURE OF ECOLOGY

1. Introduction. *Ecology* is derived from the Greek wor
"*οικος*" (oikos) meaning "house" and is the study of organism
in relation to their environment. This includes their relation
ships with other living organisms and with the non-livin
components of their environment. *Plant ecology* is the study o
plants in their natural environment but in practice it is difficul
to consider plant ecology in isolation from animal ecology
Ecology can be studied at different levels of organisation such
as the *ecosystem*, the *community* and the *habitat* (*see* **2–4** below
and when studying a particular plant one is concerned with
both the *physical* and *biotic* (i.e. biological) factors in it
environment which affect it (*see* **7–9** below).

2. The ecosystem. The most important unit of study in
ecology is the *ecosystem* which is defined as a specific part o
the earth's surface, including the plants, animals and micro
organisms in that area and all the physical aspects such a
rocks, soils, water and air as well as climate. It can be of an
size from a pond, a rotting tree trunk or a field to an island o
even a continent. The larger the ecosystem, the more difficul
it usually is to understand how it works, but even a smal
ecosystem can be extremely complex. A small temporary poc
of water may have many species of bacteria, algae and protozo
in it, together with a few higher plants and animals, and it wi
be affected by surrounding plants, animals and physica
features and by the climate of the area.

3. Communities and populations. All the living organism
within an ecosystem make up a *community*, and the study o
their relationships is termed *community ecology*. A communit
in turn is made up of *populations* of organisms, a populatio
being the members of a species of an organism in a community

Population ecology is mainly concerned with fluctuations in numbers of an organism and has concentrated on animal populations. This is because animal populations are frequently fairly mobile and there can be rapid changes in population numbers. Plants, on the other hand, live in close proximity to each other. Interrelationships between different plants are therefore often important and are studied by community ecologists who frequently pay considerable attention to vegetation patterns.

We can say that the three levels of ecological study are the population, the community and the ecosystem. An ecologist may study a population of a particular species of plant in a woodland habitat and will try to determine how that population will vary in size and distribution at different times. That will probably involve community studies in any attempt to see how the whole community interacts, but in order to understand all the aspects of the community, a consideration of the various physical factors in the whole ecosystem will be necessary. Such a study becomes very complex but, ultimately, the ecologist will seek to understand how and why the woodland is changing and will try to predict future changes.

4. The biosphere. The whole global ecosystem is known as the *biosphere* and in a sense, it is the thin "film" of life which permeates the surface layers of the planet including the atmosphere, the hydrosphere (lakes, rivers and oceans) and the lithosphere (the earth's crust). It can extend down 9 000 metres below the surface of the oceans and upwards in mountainous areas to over 6 000 metres. Above the biosphere is the *parabiosphere* where dormant forms of life such as pollen grains and fungus spores survive.

5. Biomes. The biosphere is divided into regions or *biomes*. A biome can occur in more than one part of the earth's surface but it has a particular kind of climate and soil. Consequently, it has typical animals and plants living there and these are suitably adapted to the environment of the biome.

Examples of biomes are tundra, cool coniferous forest, temperate deciduous forest, tropical rain forest and desert. The natural vegetation in Britain is mostly temperate deciduous forest, typically oak woodland, although higher and more northerly parts of Britain are naturally cool coniferous forest.

6. Synecology and autecology. An alternative way of classifying levels of ecological study is to divide it into *synecology* and *autecology*, the former being concerned with relations between species and the latter with the study of one species. This classification is now rarely used.

PHYSICAL FACTORS

7. Introduction. The kinds of plants to be found in a particular area will depend on a number of ecological factors. These are commonly grouped into two kinds:

(*a*) *physical factors* which include all those factors not directly concerned with the presence of other organisms;

(*b*) *biotic factors* which relate to the effects of other organisms.

The borderline between the two kinds of factors is not always clear because one organism will frequently affect another by having an effect on a physical factor which in turn affects the other organism; one plant, for example, may experience loss of sunlight through shading by another plant.

8. Important physical factors.

(*a*) *Light.* Plants are autotrophic, synthesising their own food from simple constituents like water, mineral salts and carbon dioxide, using the energy of sunlight. Therefore light is a key physical factor in determining the constituents of a community of plants. The demands of plants for light for photosynthesis vary considerably.

(*i*) *Shade-tolerant plants* can survive and grow in poor light conditions; many plants found growing at ground level in woodland are shade tolerant.

(*ii*) *Shade-intolerant plants* include most forest tree species, and the seedlings of such species will often only grow in forest clearings.

Because green plants require light, there is very little growth of plants below the surface layers of lakes and seas. Even in the clearest seas there will be no growth below a depth of 200 metres and most of the plants found more than a few metres below the surface are the red algae (*Rhodophyceae*) which are adapted to grow under poor light conditions. Light is also an important ecological factor in the life history of many plants, whether heterotrophic or autotrophic. It can determine processes such as the formation of spores in fungi, germination of seeds and leaf fall, with variations in day-length being crucial in many cases rather than actual light intensity. Apart from ensuring that plants flower and lose their leaves at the correct time of year, day-length can also ensure that plants of a particular species all flower at the same time.

(*b*) *Temperature*. Every plant has a range of temperatures over which it can grow, and usually a wider range of temperatures over which it may survive, even in dormant form. This range is determined partly by the nature of the plant's metabolic processes and partly by protection afforded by the plant's structure.

(*i*) *Thermophiles* are plants adapted to high temperatures and include algae that can live in hot springs.

(*ii*) *Psychrophiles* are cold-tolerant plants and include the many species of plants able to grow at low temperatures and to survive periods of freezing without becoming dormant. Many plants can survive cold periods only when in a dormant state and if subjected to cold during their active growing period will be damaged.

As well as extremes of temperature, duration of particular ranges of temperature are important to plants. Plants frequently need a minimum period of time with average temperatures above a certain level in order to grow, flower and seed. This growing season is important in determining which crops or varieties of crops can be grown under particular climatic conditions.

(*c*) *Water*. Rainfall is important in determining the constituents of a habitat, but other factors such as the distribution of rainfall during the year, the surface run-off, evaporation and the water-holding capacity of the soil all affect the availability of water to a plant.

Essentially a plant needs an adequate water supply during its periods of growth and reproduction and there are numerous modifications enabling plants to withstand drought or excessive water. On a global scale, wet areas tend to be forests, moderately dry areas are grasslands and dry areas are deserts with restricted numbers of plants.

On the basis of water availability, plants are classified under three headings.

(*i*) *Xerophytes* are drought-tolerant and inhabit deserts, semi-deserts and other dry places like bare rock surfaces where rainfall washes off quickly. Characteristics typical of xerophytic plants include very fleshy leaves covered by thick waxy cuticles in which water is stored, sunken stomata to minimise water loss, and extensive systems of thin roots to acquire water. Some xerophytes have small leathery leaves and certain xerophytic grasses roll their leaves to prevent water loss from a large surface area.

(*ii*) *Hydrophytes* are water-tolerant and usually have very diffuse tissues with plenty of air spaces to allow exchange of gases with the air. There is usually a reduced structural support system in submerged or partially submerged plants as there is less weight to support, and fully submerged plants have thin ribbon-like leaves with a large surface area relative to volume, allowing the plants to exchange gases with the water. Hydrophytes may lack root systems or they may be poorly developed and used for anchorage rather than water and mineral salt collection.

(*iii*) *Mesophytes*. Most plants are mesophytes and their characteristics have been described in VI, **11–23**. Mesophytes are harmed by inadequate or excessive supplies of water.

(*d*) *Humidity*. The amount of water in the atmosphere affects plants by affecting the drying effect of the atmosphere and hence the rate of water loss from plant surfaces.

(*e*) *Air Currents*. These affect plants in three main ways.

(*i*) *Water loss*. They affect the rate of water loss from a plant's surface.

(*ii*) *Damage*. Strong winds will damage the aerial parts of plants unless they are adapted to resist the effects of wind.

(*iii*) *Seed dispersal*. Air currents are often important in the dispersal of spores and seeds.

In aquatic plants, water currents have similar effects to air currents. In both aquatic and terrestrial plants, species may grow at a slower rate and develop dwarf forms under conditions of persistent winds or water currents.

(*f*) *Mineral elements*. Particular habitats may be deficient in any of a wide range of mineral elements required by plants for growth and reproduction. Most of these elements are required in very small amounts and may be termed trace elements (*see* VII, 22). They include iron, copper, manganese, calcium and magnesium, and plants vary greatly in their capacity to withstand deficiencies of these elements, some plants having adopted a carnivorous habit to overcome mineral element deficiencies.

Less common, except in very acid soils or as a result of pollution, is an excess of mineral elements, such as copper and lead, causing *toxicity* to plants (*see* XII, 15). Allied to mineral availability are *salinity* and *soil pH.*

(*i*) *Salinity* is most commonly found in coastal fringes such as salt marshes, the salt-tolerant plants found there being termed *halophytes.*

(*ii*) *Soil pH* greatly influences the kinds of plants found in a habitat; most plants have a preference for a particular pH range and a chalk downland, for example, will have a different community of plants to an acid moorland.

BIOTIC FACTORS

9. The nature of biotic factors. A particular organism in a habitat will be affected not just by the many physical factors but by other organisms, both plant and animal, with which it comes into contact. The three most important of these biotic factors for plants are *competition*, *predation* and the *effects of human activity.*

(*a*) *Competition.* A plant competes with its neighbours for many requirements for life including light, water and mineral salts. The closer the requirements of surrounding plants are to the plant itself, the more severe will be the competition. There are two forms of competition:

(*i*) *intra-specific competition* takes place between individuals of the same species;

(*ii*) *inter-specific competition* occurs between individuals of different species.

The most important factors for which plants compete vary with habitat. Thus in forests, light may be the most important, in deserts water is the key factor and in grassland areas availability of mineral nutrients in the soil may be the major subject for competition.

(*b*) *Predation.* Most plants are autotrophic and in making their own food they serve as a food source for the animal kingdom and part of the plant kingdom (e.g. parasitic and saprophytic plants). The ability of a plant to survive in a habitat will therefore partly depend on the extent of predation experienced in that habitat and its ability to survive attacks by organisms as diverse as parasitic viruses, bacteria and fungi as well as predatory insects and higher animals.

(*c*) *Human activity.* There has been a massive increase in the impact of human activity on plant communities especially in the last hundred years with the development of widespread industrialised societies and with large increases in the world's human population. Very few plant communities have been unaffected by human activity: many areas of fertile land are farmed to produce food, natural forests are either cleared for agriculture or exploited for timber, grasslands are grazed by domesticated animals and even remote and inhospitable regions are likely to be affected slightly by globally distributed pollutants in the oceans and atmosphere.

The effects of human activity are not always positive, and the formation of deserts, soil erosion and loss of fertility have frequently followed exploitation of natural plant communities. A reaction to this has been an increased awareness of the need to conserve natural plant resources (*see* XII, 14–22).

SUCCESSION

10. Plant communities and succession. A plant community is rarely stable and is usually developing new characteristics with new species replacing older groups of plants in a particular area. The basic reasons for this change can include changes in climate, soil fertility and even changes in the environment brought about by the plants themselves. Even if there is virtually no change in the external environment of a community, there is often a tendency for the community to evolve from simple to more complex forms, the latter having greater variability and often having a greater ability to survive changes in the environment.

11. Colonisation of rocky ground. As an example of this succession of plant species in a given environment, we can look at an area of bare rock. Initially this will be colonised by lichens, able to grow on open rock surfaces in the absence of any soil. The lichens themselves will eventually produce soil in crevices in the rock, partly by speeding up the weathering of the rock and partly by themselves growing, declining and decaying to produce organic matter. With limited amounts of soil, mosses will move in and may be more successful than the lichens in colonising the pockets of soil actually created by the lichens.

The mosses will grow faster than the lichens and will speed up the process of soil formation and eventually there will be enough soil to support higher plants such as grasses. Over many thousands of years the originally bare rock surface will become covered in soil held in place by plant roots and it will eventually become thick enough to support shrubs and even trees.

Finally, a *climax vegetation* covers the area and this will remain indefinitely unless external forces like a change in climate cause it to change. If the bare rock surface was in lowland Britain, the most likely climax vegetation would be oak woodland, with oak the *dominant* species but with many other trees, shrubs and grasses making up the complex and stable community. Fire might destroy the forest and although the bare ground would not return to a lichen colonisation, mosses would probably be the initial recolonisers and the process of succession would start again. Alternatively, a change to a colder climate would lead to the oak forest being replaced, over many years, by pine trees better able to adapt to the new climate. Biomes (*see* 5 above) represent the natural or climax forms of vegetation for different areas of the world.

ENERGY AND MATTER IN ECOSYSTEMS

12. Energy is defined as the capacity to do work and the ultimate source of energy for living organisms is the sun, utilised by green plants. These are the *producers* in an ecosystem and provide the food for the *consumers*, heterotrophic organisms, mostly animals, which consume the plants directly or indirectly. There is a third major nutritional group, the *decomposers*, and these break down the dead remains of plants and animals. In practice the situation is more complex with several levels of consumers all ultimately dependent on green plants as producers. This may be shown as in Fig. 57.

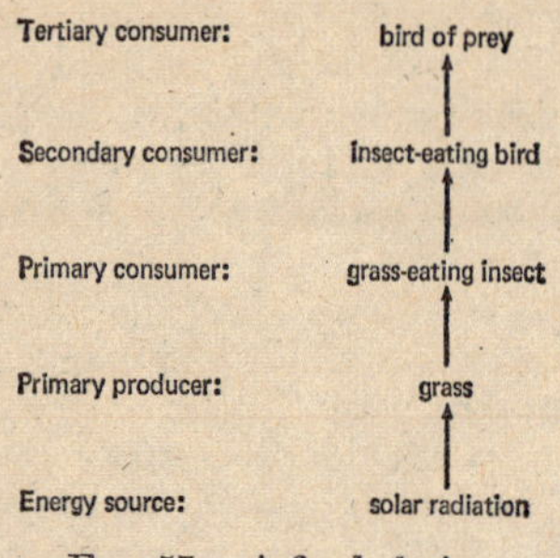

FIG. 57.—A food chain.

Even this is a simplification as all the organisms mentioned eventually decay and decomposers break down their constituents into simple chemicals which can be used by the grass and other primary producers. This means that although there is a permanent energy source (the sun) and energy flows through an ecosystem, eventually being lost as waste heat, *matter* has to be recycled, and decomposers are important in doing this.

13. Matter. This is normally cycled between organisms and the physical environment in complex processes known as *biogeochemical cycles*. Figures 58 and 59 illustrate two of the most important of these, the nitrogen cycle and the carbon cycle.

At some stage in each cycle, the element is likely to occur in an inorganic form. For example, carbon occurs as carbon dioxide in the atmosphere and nitrogen occurs as nitrites and nitrates in the soil and as nitrogen in the atmosphere. But in each case there is a stage in the cycle where the element in an inorganic form is "fixed" by a chemical process into an organic form and cycles through the living components of the ecosystem, eventually passing back into an inorganic form, often through the activities of decomposers. On a global scale, biogeochemical cycles involve massive quantities of matter but even so, human activity can have an impact on the cycles.

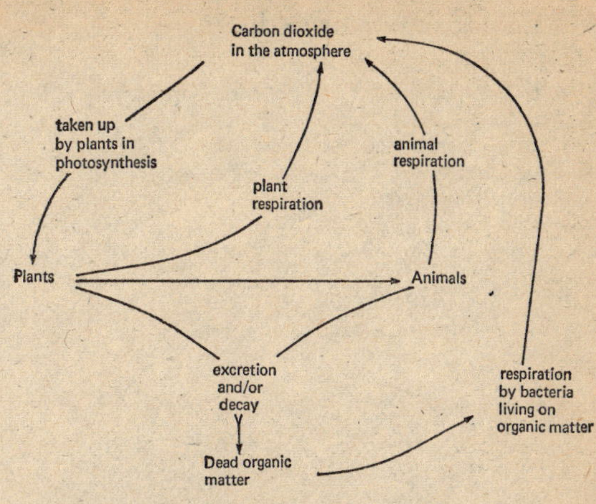

FIG. 58.—The carbon cycle.

With the carbon cycle, for example, the concentration of carbon dioxide in the atmosphere is rising very slowly (by rather more than 0·2 per cent per year) as a result of the combustion of fossil fuels.

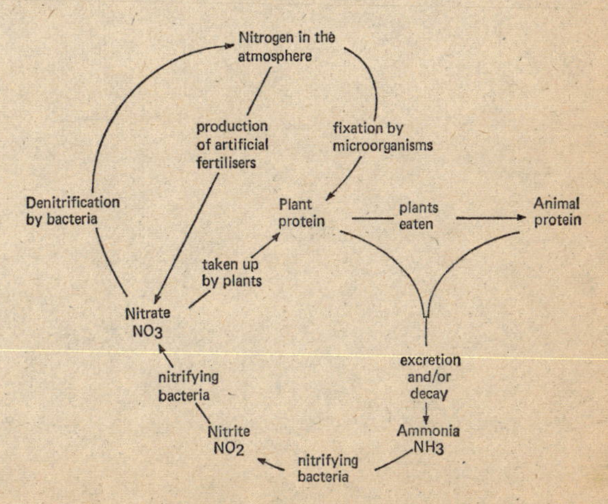

FIG. 59.—The nitrogen cycle.

PROGRESS TEST 9

1. What is an ecosystem? (2)
2. What is the difference between a community and a population? (3)
3. Give three examples of biomes. (5)
4. List four important physical factors in ecology. (8)
5. What is a xerophyte? (8)
6. Name two forms of competition. (9)
7. List some of the effects of human activity on natural ecosystems. (9)
8. What is succession? (10)
9. Give an example of a biogeochemical cycle. (13)

CHAPTER X

WORLD CROPS

THE ORIGINS OF CROPS

1. Introduction. All animals require food as a source of energy and as a source of materials for growth and reproduction. Ultimately the food that animals eat comes from green plants, made principally by the process of photosynthesis using solar energy. This applies just as much to man, but in addition to providing him with food plants also provide a wide range of other valuable resources.

2. Origin of crops. The first tool-using ancestors of modern man were probably primates such as *Proconsul*, living 30 000 000 years ago, but it was only 3 000 000 years ago that early man learnt to make tools.

Even with this ability the human population of the world remained very small until about 12 000 years ago when man first learnt to farm. This was because, before he became a farmer, man was dependent for his food on those animals he could hunt and kill, and those roots, fruits, grubs and other foods that he could gather. He had a limited food supply, food shortages were common, and although man was present throughout the world, the total population 12 000 years ago was probably less than 5 000 000.

3. Hearths of domestication. Between 12 000 and 6 000 years ago, the world's population increased about 16-fold to 80 000 000, mainly as a result of man's new-found ability to produce his own food by farming. This new skill was learnt in at least three quite separate parts of the world, the so-called *hearths of domestication*, in the Middle East, the plains around the Bay of Bengal, and in the Andean Highlands.

In the Middle Eastern hearth of domestication, wild barley (*Hordeum spontaneum*) and wild wheat (two species of *Triticum*) were domesticated along with several legume (bean and pea) crops. Grains of edible wild grasses brought back to camps probably germinated around the camps and the resulting grains were collected and eaten. In time man found it better to clear land adjacent to his camps and deliberately plant the grains. This was the start of arable farming.

4. Domestication of animals. The first animals to be domesticated included the wild Bezoar goat, the Urial sheep and wild long-horned cattle. Animal domestication probably started with the tethering of young animals after their parents had been killed by hunters. As they matured these young animals became accustomed to man and would graze in the vicinity of his camps.

As well as the three main hearths of domestication, some domestication took place elsewhere in the world, notably in West Africa and parts of China. Over many thousands of years plants and animals which had become domesticated were spread around the world as part of human migrations and it is now often difficult to tell the exact origins of the many widely grown plants cultivated by man.

FOOD CROPS

5. Cereals. On a world scale, the most important cereal crops are *wheat*, *rice* and *maize*. Common but less important are *barley*, *oats*, *rye*, *sorghum* and the *millets* (*see* **6–10** below).

6. Wheat. There are many species of wheat, all originating in the old world and all important as crops of temperate regions. The species are grouped into three categories depending on whether they have fourteen, twenty-eight or forty-two chromosomes. Of these the fourteen chromosome or diploid wheats are probably the most ancient, including two species of "einkorn" wheats, so-called because each spikelet has one seed. Wheat of this category is still grown in parts of Europe and Western Asia but is used mainly as a whole grain for feeding cattle and horses rather than for bread-making.

The *tetraploid* or twenty-eight chromosome wheats of the second group probably originated in the crossing of wild einkorn with another grass to produce wild emmer, a hybrid found naturally in the Middle East. The cultivated form of this plant has been found in ancient archaeological sites in Iraq and may have been cultivated even earlier than einkorn wheat.

Finally, the *hexaploid* or forty-two chromosome wheats are the ones which have been developed most for modern wheat growing; indeed they do not occur naturally. The hexaploid wheats have certainly been grown for 4 500 years and their value lies in their productivity and their content of *gluten*, the protein that gives bread its texture.

Cultivated wheats can, for the most part, be divided into winter and spring wheats, the former being planted before the onset of winter and the latter, relatively fast-growing, being planted in spring and taking about 100 days to grow and mature.

7. Rice. This is derived from wild grasses of the genus *Oryza*, there being over twenty-five wild and cultivated species. It is the staple food of monsoon Asia, but is becoming increasingly important in Africa and America. It apparently originated in South-East Asia and was in use at least 5 000 years ago and probably much longer.

There are two main ways of growing rice, either in dry ground or in watery fields. Most rice is grown in the latter way and is termed paddy as distinct from upland rice. Paddy rice is sown in rich fine soils and transplanted to paddies after 40 days. The paddies are drained as the rice ripens and after some three to nine months, depending on variety and climate, the rice is harvested. Intensive efforts over many years have resulted in the production of high-yielding rice varieties, especially at the International Rice Research Institute in the Philippines. These new varieties helped to form the basis of what has been called the "green revolution", the introduction of new high-yielding varieties of many cereal and legume crops in tropical countries.

8. Maize. This is a new world crop but has now spread throughout the world and is the staple food of many people in Africa. Wild maize is unknown and it was probably first cultivated many thousands of years ago in South America. *Zea mays* is a tall annual grass with male and female flowers, the latter eventually producing the maize cob. Unlike many cereals, maize does not tiller (i.e. it does not produce more than one shoot from a seed) but it can vary in height from 1 to 6 metres depending on the variety. The cob can be from 20 to 60 cm in length with many rows of closely packed seeds on each cob. The many different varieties include forms which produce sweet corn, flour corn with a soft mealy grain and corn suitable for cattle feeding. Along with sorghum (*see* **10** below), maize is frequently termed a coarse cereal grain, unlike the other cereal crops known as small grains.

9. Barley, rye and oats. These are all temperate cereals, barley being the most important. It is generally a hardy crop and is used primarily as a cattle feed and for producing malt as a precursor of beer. Oats are used mainly for animal feeding and are also hardy when compared with most wheat varieties. Rye is relatively unimportant although it retains its use as a constituent of breads in some parts of the world.

10. Other cereals. Many other cereals are grown, especially in the tropics. These include sorghum, a cereal similar in size to maize but producing more diffuse seed heads widely used for making a coarse flour and for fermenting into beer. One useful attribute of sorghum is that it is more resistant to drought than maize. The millets vary in size from the small finger millets to the much larger bull-rush millet, the latter being similar in size to sorghum. They are common in many parts of Africa.

11. Sugar. This is derived from two main plants, *sugar cane* and *sugar beet.*

(*a*) *Sugar cane* (*Saccharum officinarum*) originated as a river bank grass in south-eastern Asia but is not now found wild, although many related species occur in the tropics. Modern sugar cane varieties have been derived from hybridisations made with many different species, but the sugar cane remains a plant of tropical and semi-tropical regions. It is one of the fastest growing crops and it is not uncommon to get up to 250 tonnes of cut cane per hectare in the space of 14 months. It is a tall grass, up to 5 metres high and the hard stem contains a very sweet sap with over 12 per cent sugar content. When harvested the cane is crushed in large mills and the juice purified, concentrated and allowed to crystalise into sucrose.

(*b*) *Sugar beet* (*Beta vulgaris*) is quite different, being a temperate root crop derived from a coastal plant (*Beta maritima*) in the past 200 years. It is biennial, not normally flowering in the first year, but in commercial sugar production the swollen roots are lifted in the first autumn and the beets are crushed to extract the sweet juices which are purified in the same way as with sugar cane. Sugar beet usually contains over 15 per cent sucrose and the crop as a whole produces about one third of the world's supply of sugar.

12. Root crops. Many plants reproduce by asexual means, often involving the formation of storage organs which can survive harsh seasons and produce new plants with the advent of subsequent growing seasons. Many of these storage organs are used by man for food and most frequently take the form of tubers, rhizomes and bulbs (*see* **13–18** below).

13. The potato (*Solanum tuberosum*) is a native of South America but is now most commonly grown in Europe, western Asia and North America. It is a very bulky crop producing up to 40 tonne of tubers per hectare and although the nutritional value per unit weight is much lower than most cereals it is still a very good source of food. It thrives in a moderately cool damp climate, too much heat in damp climates making it succumb to many rotting diseases. It grows vigorously during summer months and, with the onset of autumn, builds up large food reserves in undergound tubers. Potatoes are planted at intervals of 40 cm in rows 60 to 90 cm apart and after a few weeks of growth, the stems are earthed up to encourage tuber formation. The many different varieties include first and second earlies which can be harvested some months before the heavier-yielding maincrop varieties usually harvested in late autumn.

14. Turnips (*Brassica rapa*) and swedes (*Brassica campestris*). These are important temperate crops, shallow rooted and both liking damp ground. Each plant forms a large single underground storage organ, the leaves of the turnip being rough and green and coming straight from the swollen root whereas in the swede they are blue-green and grow from a neck of tissue on the swollen root. The *mangold* (*Beta vulgaris*) is more cylindrical than the turnip or swede and can tolerate drier conditions. It is used exclusively as an animal feed whereas the other two crops are eaten by man as well.

15. The onion (*Allium cepa*) is a biennial belonging to the *Liliales*, as do its close relatives the *garlic*, *leek* and *shallot*. They are bulb crops developing a bud surrounded by thick fleshy white leaves and can be grown either from seed or from small bulbs. Their food value is relatively low but they are prized for their flavour which is considered to accentuate the flavour of other foods.

16. Tropical root crops. Many tropical root crops are very important in subsistence agriculture where farmers grow enough food just for themselves and their families. Subsistence agriculture utilises 40 per cent of the total land area of the world devoted to agriculture and feeds more than 50 per cent of the world's population.

17. The cassava (*Manihot esculenta*) is also known as the tapioca or manioc plant and is grown throughout the tropics producing huge fleshy tubers harvested between 15 and 2[illegible] months after planting. Cassava is a large plant about [illegible] metres high with a spreading canopy. It is not very tasty and has to be cooked or dried in the sun to denature the hydrocyanic acid which it contains. It is either turned into a flour or cooked and eaten like a potato and one of its important qualities is its drought resistance.

18. The yam (*Dioscorea spp.*) is a climbing plant which likes a wet climate and produces a massive swollen root, often 30 cm long and 15 cm across. It takes nearly a year to reach maturity but, unlike the cassava, stores quite well. The *sweet potato* (*Ipomoea batatas*) and *taro* (*Colocasia antiquorum*) are both common tropical root crops, the former being cultivated much like the Irish potato and the latter producing very deep-seated corms, usually in very wet soil.

19. Greens. There are many hundreds of species of plants grown for human consumption of the leaves. An important group belongs to the genus *Brassica* and includes the *cabbage*. The plant is biennial, making leaf growth in the first year and flowering in the second. It is therefore harvested in the first year and varieties are available which can be planted at almost any time of year. The leaves are normally boiled before eating as are the miniature heads of the *Brussels sprout*, related to the cabbage but producing many buds in leaf axils up the stem. Related also to the cabbage are *cauliflowers* and the similar but hardier *broccoli*, the difference being that the flower heads rather than the leaves of these plants are eaten. Many brassicas such as *kohlrabi* and *kale* are grown mainly for animal feed.

Although not related to the brassicas, *lettuce* (*Lactuca sativa*) is a common and popular leafy vegetable grown in many parts of the world as a constituent of salads. The *rhubarb* plant (*Rheum rhaponticum*) is one of the hardiest of all crops, originating in cold parts of Asia and producing succulent reddish stems early in the growing season. The leaves are not edible due to a high concentration of oxalic acid.

20. Seed crops. The most important of the seed crops are the peas, beans and related legumes, all members of the *Leguminosae*. Their importance stems from two attributes:

(*a*) *They are often very rich in proteins and oils;*

(*b*) *their roots have nodules containing symbiotic bacteria* which, in co-operation with the plants, fix atmospheric nitrogen and increase the nitrogen content of soil.

In temperate regions the *green pea* (*Pisum sativum*), *runner bean* (*Phaseolus multiflorus*) and the more hardy *broad bean* are all important vegetables. The *haricot bean* (*P. vulgaris*) is processed to produce "baked beans" especially popular in Britain.

Among the hundreds of tropical and sub-tropical legume crops, the two most important are the *soy-bean* (*Glycine soja*) and the *ground-nut* or *peanut* (*Arachis hypogaea*). The former has a very high protein content, although rather deficient in the sulphur-containing amino acids. It is much prized for its oil and the dried meal remaining after extraction of the oil is used as a high-protein animal feed. The ground-nut produces yellow flowers which grow down into the ground producing the hard-coated seeds underground, hence the name. It is grown in many tropical countries for export as a cash crop to temperate countries.

21. Fruits. Fruit crops can be divided into soft- and hard-stemmed plants and there are many species of each category grown in tropical and temperate regions.

Important temperate soft-stemmed fruits are *tomatoes* (*Lycopersicum esculentum*), *cucumbers*, *marrows*, *squashes* and *pumpkins*, grown under glass in colder regions. Biennial and perennial soft-stemmed fruits include *strawberries* (*Fragaria spp.*), *blackberries* and *raspberries* (both *Rubus spp.*) In the raspberry, perennial underground roots send up leafy stems each summer which are biennial, producing fruits the following year.

Hard-stemmed temperate fruits include many species of the *Rosaceae* (rose family), mostly long-lived perennial trees cultivated in orchards. Most ripen in late summer or early autumn and include *apples* (*Malus sylvestris*), *pears* (*Pyrus communis*), *plums* (*Prunus domestica*), *cherry* (*Prunus avium*) and the less hardy *peach* (*Prunus persica*) and *apricot* (*Prunus armeniaca*).

Soft-stemmed tropical fruits include many species from temperate climates which grow well under tropical conditions. In addition, there are less hardy plants, some of the largest being the *dessert* and *cooking bananas* (also termed *plantains*). The former are grown as cash crops, often for export, whereas the latter are subsistence crops, being picked, and then cooked and eaten like the potato and having very little "banana" flavour. The trunk of a banana is really a succession of leaf bases produced from an underground rhizome. It is often 7 metres high with 4 metre long leaves taking 18 to 36 months to produce large bunches of fruit weighing up to 50 kg a bunch. Bananas need a high rainfall, more so than the *pineapple* (*Ananus spp.*), a plant native of arid grasslands of South America. Well adapted to resist grazing animals by having sharp spikey leaves, it produces a fruit which is actually composed of many fruits around a swollen stem.

Among primarily tropical and sub-tropical hard-stemmed fruits, the citrus fruits are most important, all plants belonging to the genus *Citrus*. *Oranges*, *grapefruit* and *lemons* all have a thick outer peel and a juicy fleshy inner pulp usually divided into about ten segments, and they all produce strongly-scented oils.

The tropical palms are the most important monocotyledonous tree crops. The *coconut palm* (*Cocos nucifera*) is a tall graceful tree up to 20 metres high with a crown of long leaves and with the fruits borne on the stem at the centre of the crown. The woody-shelled nut is enclosed in a thick fibrous coat, called the coir, with a green skin. The crop has many uses:

(*a*) *The leaves* are used for floor and roof coverings;

(*b*) *The coir fibres* are made into baskets, mats and ropes;

(*c*) *The flesh and milk* are eaten;

(*d*) *The milk* is fermented to make a strongly alcoholic drink from which the spirit *arrack* is distilled;

(*e*) *Coconut oil* is extracted from the coconut flesh.

The *date palm* (*Phoenix dactylifera*) grows well in dry areas provided some water is available from underground supplies. The tree produces dates within four years of planting and when fully mature will produce 50 kg of dates a year and these keep well if dried.

The third important but less well known palm tree is the *oil palm* (*Elaeis guineensis*), a short squat tree which gets thicker at its base as it ages. Large bunches of fruit are produced, each consisting of an oil-rich seed in a thick pulpy coat. Wild fruit are gathered in many parts of Africa but the palm is most commonly grown on large plantations, the oil that is extracted from the seed being used for making margarine, cooking fat and lubricants.

22. Pasture crops. Of great importance but usually neglected in economic botany are the many plants grown for pasturing domesticated animals. Pastures are either permanent or ley, the latter being seeded pastures, often part of a rotation, and the major constituents are grasses and legumes, the important ones being listed below:

(*a*) *Grasses*: Italian rye grass (*Lolium italicum*); perennial rye grass (*L. perenne*); timothy (*Phleum pratense*); meadow fescue (*Festuca pratensis*); cocksfoot (*Dactylis glomerata*).

(*b*) *Legumes*: white clover (*Trifolium repens*); broad red clover (*T. pratense*); alfalfa or lucerne (*Medicago sativa*).

OTHER USES OF PLANTS

23. Introduction. Although plants are most important to man as a source of food, they have a very wide range of other uses. Plant products are used to make fibre goods, perfumes drugs, oils, dyes, fuel and paper, and wood is used for many purposes. Some of the main non-food uses are described in 24–30 below.

24. Perfumes and flavours. All the spices such as *cinnamon* *vanilla*, *ginger*, *mustard*, *peppers* and *chillies* are plant products frequently dried powders prepared from roots, flowers or fruits. Most are derived from tropical crops and the spice trade between the Far East and Europe was one of the principal reasons for the establishment of communications between these two regions of the world. Although these flavours and herbs are consumed, their food value is limited and they are seldom classed as food crops.

Important beverages include *tea*, produced from the leaves of the bush *Camellia sinensis*, *coffee*, produced mainly from the beans of the *Coffea arabica* and *C. robusta* trees, and *cocoa* produced from the many beans found in each large pod of the *Theobroma cacao* tree. This last tree is an unusual plant in that flowers and subsequently fruit pods are produced on small branches coming straight out of large trunks. All three crops are very important cash crops exported mainly from less developed countries. Coffee was, until recently, the second most important commodity on world markets after oil.

Perfumes are produced from a very wide range of plants, two of the most important being *lavender* (*Lavendula officinalis*) and the *damask-rose*.

Analogous to the tropical spices are the temperate herbs long used in flavouring and also important medicinally. They include *sage* (*Salvia officinalis*), *thyme* (*Thymus vulgaris*) and *peppermint* (*Mentha piperita*).

25. Drugs. *Tobacco* (*Nicotiana tabacum*) may be correctly listed in this category, the principal active ingredient of the dried leaves of the plant being nicotine. Drugs obtained originally from plants although sometimes produced by industrial synthetic processes include *morphine*, *quinine*, *digitalis* *cannabis*, *atropine*, *strychnine* and *cocaine*. The important *pyrethrin* insecticides come from the *Pyrethrum* flower, grown commercially in Kenya. Strictly speaking, even the antibiotics such as *penicillin* and *streptomycin* are produced by actinomycetes, fungi and other primitive plants.

26. Oils. Many food crops including maize, ground-nut, soybean and coconut are also grown for vegetable oil production, and *castor oil*, *linseed*, *olive* and *sun flower* are examples of crops grown almost entirely for their oils. Almost all of these oils have many uses including the production of margarine, candles, soap, lubricants and, in the case of castor oil, for medicinal purposes.

27. Dyes. Although the majority of dyes are now produced by industrial synthetic processes, plant dyes are still used by some craftsmen and tan-stuffs are still obtained from plants, usually in the form of extracts from trees such as the *wattle* (*Acacia spp.*).

28. Fibres. Of the important plant fibres, *cotton* is the best known. It is produced from the seed hairs of an annual plant (*Gossypium spp.*) and is grown in both sub-tropical and tropical countries. Competition from synthetic fibres has not succeeded in displacing cotton as a major fibre. The waist-high plant thrives best in a climate which combines wet growing weather followed by a dry season for the crop to mature, this combination producing high quality fibres.

Sisal (*Agave sisalana*) bears an array of very thick tough spikey leaves at ground level surmounted by a spectacular flowering stalk 5 to 7 metres high, the fibre being obtained from the leaves and being used mainly in making string and sacks. *Jute* (*Corchorus capsularis*), by comparison, is a tall slender plant in which the stem fibres are important and the leaves are of no value. Both are tropical crops but jute likes a wetter climate. It is harvested when about 3 metres high and is used for making sacking and the fibre backing for floor coverings.

Hemp (*Cannabis sativa*) and *flax* (*Linum usitatissimum*) are tropical and temperate fibre crops respectively, the latter being the source of linen. The former produces the drug cannabis under dry conditions, but needs a wet climate for good quality fibre production.

29. Cork and rubber. *Cork* comes primarily from the bark of the *cork-oak* (*Quercus suber*) in Portugal, Spain and Algeria. *Rubber*, a commodity of major world importance, can be produced synthetically, but a large proportion still comes from plantations of the *rubber tree* (*Hevea brasiliensis*) in West Africa, South America and especially South-East Asia. It is produced from latex, a milky white fluid produced by the tree to seal wounds. The bark of each rubber tree is carefully tapped to extract latex about twice a week and provided the operation is done carefully, a tree will continue to yield for many years.

30. Timber and paper. There are two main sources of timber, one of which is also the main source of paper.

(*a*) *Softwoods*. These are woods of coniferous forests especially those girdling the northern hemisphere through Scandinavia, Russia and North America. The most important softwood trees are *Sitka spruce* (*Picea sitchensis*) and *Douglas fir* (*Pseudotsuga menziessi*) and softwoods such as these are sources of pulp in paper making and are also used for construction.

(*b*) *Hardwoods*. These are used primarily for construction and for furniture-making. There are important temperate and tropical species.

(*i*) *Temperate hardwoods* include *oak* (*Quercus spp.*) which is the most valuable for construction purposes, *beech* (*Fagus spp.*) and *sycamore* (*Acer spp.*) useful for furniture-making and *ash* (*Fraxinus spp.*) used for making handles.

(*ii*) *Tropical hardwoods* include teak (*Tectona grandis*) sapele (*Entandrophragma cylindricum*) from tropical Africa and Honduras mahogany (*Swietenia macrophylla*).

PROGRESS TEST 10

1. How long ago did man first learn how to farm? (2)
2. What is a hearth of domestication? (3)
3. What are the world's most important cereal crops? (5)
4. What are the two ways of growing rice? (7)
5. What are the main uses for barley? (9)
6. What are the two sources of sugar and where are they grown? (11)
7. List four root crops. (13, 14)
8. What is a yam? (18)
9. Describe two important legumes. (20)
10. Why are legumes important? (20)
11. What are the uses of the coconut palm? (21)

CHAPTER XI

ECONOMIC BOTANY: WORLD FOOD SUPPLY

WORLD FOOD PROBLEMS

1. Introduction. One of the major world problems is finding the means to feed the growing population of the world. If it is to be solved, then most of the emphasis will have to be on improving the production of food by growing more and better crop plants.

2. Malnutrition. In 1974 there were estimated to be 460 000 000 people throughout the world who were not getting sufficient food to lead a normal life. Many were suffering from deficiencies of the major food groups, carbohydrates, proteins and fats, and many more from deficiencies in vitamins and minerals. The effects of this malnutrition included decreased resistance to disease, a low rate of growth, lethargy, mental retardation and, in extreme cases, starvation and death.

The world's population is growing rapidly and according to the Food and Agriculture Organisation (F.A.O.) of the United Nations, there may be as many as 750 000 000 malnourished people in the world by 1985 unless more effort is devoted to improving world food supplies. In order to ensure adequate food supplies and to avoid malnutrition, the F.A.O. believes that world food production in 1985 should be 45 per cent higher than in 1970.

The areas of the world where malnutrition is most common are southern and south-east Asia, large parts of the Middle East, most of Africa, Central America and the northern parts of South America.

3. The green revolution. The less developed or poorer countries of the world, mostly in the tropics, are the countries where malnutrition is most common, but during the 1960s these countries generally managed to increase food production sufficiently to keep pace with their growing populations. Part of the reason for this success was the "green revolution", the development by plant breeders of new varieties of tropical crops which produced far more food per unit area than the old varieties. Two international research stations partly responsible for this work were the International Maize and Wheat Improvement Centre in Mexico and the International Rice Research Institute in the Philippines. The latter (I.R.R.I.) produced two of the best known crop varieties in the green revolution, two new varieties of rice known as IR8 and IR11.

4. The world food crisis 1973–5. Although the green revolution was initially partially successful, the world food situation deteriorated suddenly and rapidly early in the

TABLE III: WORLD FOOD SECURITY INDEX[1]

Year	*Reserve stocks*	*Stock equivalent of idle land*[2]	*Total reserves*
1965	147	71	218
1966	151	78	229
1967	115	51	166
1968	144	61	205
1969	159	73	232
1970	188	71	259
1971	168	41	209
1972	130	78	208
1973	148	24	172
1974	108	—	108
1975	111	—	111
1976	100	—	100
1977	180	—	180

NOTES:

1. All figures are in millions of tonnes of grain (Source: World-watch, Washington).
2. In most years a large area of grain land in North America is kept idle and is only planted when there is sufficient demand. When idle this represents an additional reserve in terms of the amount of grain which could be grown if need be. There was no idle land in 1974–7.

1970s. The effects of this deterioration included famines in many tropical countries, especially in Africa and Asia, as well as a decline in the world reserves of grain. The accompanying Table (Table III) giving the World Food Security Index shows how quickly world grain stocks declined in 1973 and 1974. There were at least six different reasons for this situation and the factors involved tended to interact to make the overall problem more difficult to solve. The major factors were as follows.

(*a*) *Population growth.* There have been increases in population in almost every country in the world in recent years and the world's population is now growing at the rate of 2 per cent each year, an annual increase of 100 000 000. This means that the population of the world will rise from the 1975 figure of 4 000 000 000 to nearly 7 000 000 000 people by the year 2000. Population growth is usually highest in the poorer parts of the world (*see* Table IV) and many individual countries have growth rates of over 3 per cent each year meaning that they can expect to double their population in about 23 years or less.

TABLE IX: REGIONAL VARIATIONS IN POPULATION GROWTH[1]

Region	*Annual rate of population growth (per cent)*	*Proportion of population under age of 15 (per cent)*
Africa	2·7	44
Latin America	2·9	42
Asia	2·3	40
North America	1·3	30
Europe	0·8	25
Oceania	2·0	32
World Average	2·0	37

1. Source: UN Data.

Apart from having to raise food production just to keep pace with population growth, countries with rapid population growth also tend to have a large proportion of their people under the age of 15 years with an especially large proportion of young children who need feeding but are too young to contribute to food production.

(*b*) *Neglect of rural development.* In the past thirty years most of the emphasis in development in the poorer countries of the world has been on urban and industrial development. This has included the building of airports, roads, railways, ports, large hospitals, colleges, factories, oil refineries and steel mills. Much of this development has benefited food production indirectly, for example, by improving communications, but it has usually been concentrated in the towns and cities. In most poor countries the majority of the population live in the rural areas and are involved in agriculture. There has been a lack of money available to develop the rural areas and especially agriculture, and the result has been that food production has barely kept pace with increased demand.

Although population growth and lack of rural development have been the long term causes of world food problems, the sudden crisis in the early 1970s came about because several short term factors had an unexpected effect on food production.

(*c*) *Bad weather conditions.* One of these was a combination of droughts in some countries with floods in other countries, each causing poor harvests. Harvests in some temperate countries including the United States and the Soviet Union were also affected by bad weather, sometimes by late frosts but also by droughts. Some poorer countries were very badly affected and the Sahel zone of central and northern Africa, including more than ten countries, experienced a 7-year-long drought ending only in 1974.

(*d*) *Fertiliser shortages.* A world-wide shortage of fertilisers developed in the early 1970s and this affected the less developed countries most as they were least able to pay the high prices caused by the shortage. This was particularly serious because many of the best high-yielding crop varieties of the green revolution only gave really good yields if they received adequate amounts of fertilisers.

(*e*) *Increased food consumption in developed countries.* There was a tendency in many rich developed countries in the 1960s and early 1970s for people to eat more meat, mainly because they were better able to do so. Much of the meat consumed in these countries, especially beef and poultry, is produced from animals which have been fed partly on grain-based foods. About 40 per cent of world grain production is used for making animal feeds, but most meat-producing animals need 7 to 12 kg of feed to make 1 kg of meat. Increases in demand for meat can therefore put a proportionally greater strain on world grain production.

(*f*) *The increase in the price of oil.* The final factor in causing world food problems was the increase in the price of crude oil in late 1973 and early 1974. Within a year the price of crude oil went up by over 400 per cent and this had a number of effects on food production, especially in less developed countries. The oil price rises led to higher prices for fuel for tractors, irrigation pumps and other items of agricultural machinery. It also meant that pesticides and fertilisers used in agriculture increased in price because they were either made from oil-based chemicals or else used a lot of energy in manufacture. Poor farmers were not able to afford the increased costs and food production suffered.

5. The 1974 World Food Congress. The United Nations Organisation responded to the world food crisis by calling a World Food Congress in Rome in 1974 which was attended by delegates from 130 countries. The congress had two main results.

(*a*) *Overcoming of grain shortage.* Attempts were made to meet a shortage of 10 000 000 million tonnes of grain affecting twenty-two countries in Asia, Africa and Latin America. While these attempts were not actually successful during the Congress, within 6 months most of the shortages had been overcome.

(*b*) *Long-term food production plan.* Probably more important was that a long-term plan was accepted at the congress which sought to establish a new World Food Council with the task of ensuring that world food production would be improved sufficiently by 1985 to meet all world food needs. In order to achieve this, a 45 per cent increase on 1970 production levels, it was estimated that spending on agricultural development would have to treble to $5 000 000 000 per annum until 1985, with most of the emphasis on increasing the food produced from plants.

IMPROVING WORLD FOOD PRODUCTION

6. Introduction. Methods available for improving world food production include bringing new land under cultivation improving cattle and fish production, growing better crops and utilising plant and animal material that is currently considered unpalatable. Because of shortage of fertile land and the cost of bringing poor land under cultivation, growing food in new areas is mainly limited to using grazing animals to graze marginal land effectively. Freshwater and marine fisheries do have potential for expansion in many parts of the world, but most of the impetus for improving world food production must come from increasing yields of crops. This is especially true in tropical countries where small farmers produce most of the food, the great majority of it by growing crops. To improve crop production requires considerable investment and the encouragement of stable markets. Given these requirements, however, there are a number of ways of increasing crop yields.

7. Plant breeding. It is normally possible to breed new varieties of crops from older varieties so that they combine the best features but do not have the disadvantages. For example, a high-yielding variety of a crop which is susceptible to a major plant disease can be crossed with a low-yielding but disease-resistant variety, and it should be possible to arrive at a new high-yielding but disease-resistant variety. If this is successful then the new variety is grown just for its seed until there is sufficient seed to distribute to farmers. The advantage of this is that the new variety should automatically produce more food than the old one. The disadvantage is that plant breeding is a highly skilled process and the selection of new varieties and the production of adequate amounts of seed for distribution is a long process taking up to 12 years.

8. Irrigation. Usually the most important factor limiting food production in tropical agriculture is lack of water. Irrigating crops, even using primitive methods, may be the most effective short-term method of improving crop yields. Successful irrigation practice requires either a perennial water source such as a dependable large river or else some means of storing water to irrigate crops in dry periods. It is sometimes possible to use underground water supplies and even some of the driest parts of the world have adequate underground water supplies.

The use of irrigation to improve food production is expensive and unless done with skill can lead to the build-up of salt in the soil. This in turn leads to *saline* soils which can prevent the growth of crops; large areas of southern Asia have been harmed by faulty irrigation systems.

9. Fertilisers. The use of artificial fertilisers containing nitrogen, phosphorus or potassium (and compound fertilisers containing more than one of these essential elements) can greatly increase crop productivity. They are, however, often expensive especially in remote rural areas where transport costs are high. Useful alternatives are the use of human and animal wastes and green manure. In the latter case, plants with root nodules which can make their own nitrogen compounds by fixing atmospheric nitrogen are grown before a food crop (*see* VII, 21). When fully grown they are ploughed into the soil and allowed to rot and improve soil fertility.

10. Pest and disease control. Pests and diseases destroy about one third of crops grown in the tropics. They include insects and fungi as the most important pests and diseases, but nematode worms, bacteria, viruses and even birds and animals are also involved. The qelea bird, for example, a kind of weaver bird, can exist in large flocks which can do as much damage to crops as a swarm of locusts.

Pests and diseases can be controlled by three main methods, *chemical*. *cultural* and *biological* control.

(*a*) *Chemical control*. This involves destroying the pest or disease before it can harm the crop and can be used in a number of different ways.

(*i*) *Eradicants* destroy pests before they reach crops, e.g. aerial spraying of a swarm of locusts;

(*ii*) *Protectants* are sprayed or dusted on to plants and coat them with a protective covering which kills the pest;

(*iii*) *Systemic* pesticides are taken up by plants and kill the pests after they have started to attack the plant.

There are two disadvantages in the use of chemicals to protect crops. One is that they are often expensive, require machinery such as sprayers and dusters for their use and also need skilful handling. The other is that pests and diseases can develop *resistance* to the pesticides and can then be unaffected by their use.

(*b*) *Cultural control* is normally far less expensive than chemical control but may not produce such good results. Two methods are *plant quarantine procedures* and *intercropping*.

(*i*) *Plant quarantine procedures* try to prevent plant pests and disease spreading from one country or even continent to another. Most crop pests and diseases are not distributed throughout the world. If the movement of plant material is controlled by customs authorities, spread of diseases and pests may be held in check. This method requires the help of the general public.

(*ii*) *Intercropping* can be used in simple agricultural systems where there is no mechanisation and it is practicable to grow plants mixed up together. It is highly effective in controlling pests and diseases because few of them attack more than one species of plant. For example, by growing soybeans, maize and sweet potatoes together, diseases of any of the crops have great difficulty in getting from one host plant to another.

(*c*) *Biological control.* The aim here is to use one organism to kill or incapacitate another organism which is a pest or disease of plants. An example was the introduction of the *Cactoblastis* moth into Australia from its native Argentina about fifty years ago in order to control the prickly pear cactus which had become a serious weed. The effective use of biological control normally requires a detailed knowledge of the life history of both the pest and the biological control agent. Its great advantage is that it requires no chemicals and its effects should spread naturally where the pest is a problem.

11. Prevention of storage losses. In addition to the food lost to pests and diseases before harvest, about 25 per cent of all food actually harvested in the tropics is lost through storage pests and diseases. The most common pests are insects, rodents, fungi and bacteria, the latter two most often causing rots. Good storage methods involve provision of clean, dry, well-ventilated stores which prevent pest access. While this is difficult, especially in humid regions, good storage is not necessarily expensive and can often be achieved using locally available materials.

12. Mechanisation. Agriculture can be mechanised by the use, for example, of tractors for ploughing, sowing, weeding and harvesting and by the use of pumps for irrigation purposes and for spraying pesticides. While considerable potential exists for mechanising tropical agriculture, lack of labour is rarely a problem and more attention is now being focused on using labour more efficiently rather than engaging in expensive mechanisation programmes. *Intermediate technology* is concerned with this approach and may involve, for example, the development of efficient animal-drawn implements and even pedal-operated mills.

13. Novel plant food sources. In addition to improving crop production, there is great potential for using new sources of plant food such as *oilseed meal* and *single cell protein*.

(*a*) *Oilseed meal.* Important oilseeds such as soybeans and ground-nuts are processed to extract the oil, leaving behind a dry protein-rich meal, usually used as cattle feed or fertiliser. This can now be processed and upgraded by the addition of other plant proteins to produce a food supplement, rich in protein, which can be added to many other foods to improve their nutritional value.

(*b*) *Single cell protein* is produced from the dead cells of micro-organisms such as bacteria, yeasts and algae grown on many different substrates including industrial waste products. Known as S.C.P., it has been used for many generations in some parts of the world, and in parts of central Africa a staple food is *dihe*, a thick protein-rich soup made from drying off the green algal growth found in brackish ponds. S.C.P. is now produced by industrial fermentation processes and the final product contains 35 to 75 per cent of protein by dry weight. In Britain it is only used as an animal feed at present. The advantages of S.C.P. are that the micro-organisms grow quickly and can feed on waste products, but the disadvantages include high capital costs and heavy energy consumption, especially in processing the S.C.P. after production.

14. Conclusions. Sections **6–13** describe some of the main ways in which world food production can be improved by the more efficient production of food from plants. All these ways, however, require financial investment. While the plant scientist can contribute greatly to ensuring a well-fed world, it will not be possible unless the resources are available for the work to be done.

PROGRESS TEST 11

1. How many people suffer from malnutrition? **(2)**
2. What is the green revolution? **(3)**
3. Why did the world food situation get worse in 1973–5? **(4)**
4. What were the achievements of the 1974 World Food Congress? **(5)**
5. How may plant breeding improve food production? **(7)**
6. What proportion of crops is lost through pests and diseases? **(10)**
7. What are the three main ways of controlling pests and diseases? **(10)**
8. What is oilseed meal? **(13)**
9. What is single cell protein? **(13)**

CHAPTER XII

APPLIED ECOLOGY

1. Pollution. A *pollutant* may be defined as a waste product of human activity which is harmful to living organisms. The harmful effects of a pollutant may vary from killing an organism to causing temporary or chronic injury. It may even include damaging the amenity value of an area, making it less enjoyable an experience to live there. Two of the most important forms of pollution are air pollution and water pollution.

AIR POLLUTION

2. Major forms of air pollution. In industrialised countries, the three major forms of air pollution are *smoke*, *sulphur dioxide* (SO_2) and *photochemical smog*. The last one is only a serious problem in large cities which have a warm, sunny climate, at least for part of the year, and also have a heavy concentration of motor vehicles.

3. Smoke and sulphur dioxide. Both of these are produced mainly by the burning of fossil fuels, with smoke produced most commonly through the burning of coal. The following table shows the changes in smoke and sulphur dioxide emissions (not concentrations in the atmosphere) in Britain in recent years, the figures being in millions of tons per year.

	1953	1968	1975 (estimate)
Smoke	2·3	0·9	0·6
Sulphur dioxide	5·0	6·0	5·4

The figures show that, in Britain at least, there was a marked decline in smoke emissions between 1953 and 1975, mainly as a result of clean air legislation. Sulphur dioxide emissions did not fall, mainly because of the greatly increased use of fossil fuels in generating electricity.

4. Local air pollution concentrations. Emissions do not indicate the local variations in air pollution. Towns, cities and industrial areas have much more air pollution than rural areas, although the building of power stations with 200 metre high chimneys means that sulphur dioxide is dispersed more widely than two decades ago. Local topography is important and towns in valleys often have higher pollution levels than normal. This is especially true if the valleys run in the direction of the prevailing winds, so that the air currents "collect" pollutants which accumulate downwind of the town.

5. Thermal inversion layers are layers of warm air which are virtually stationary and overlie colder air at ground level. Such layers prevent the escape of air pollutants which can accumulate at ground level. In the December 1952 smog which affected London, smoke and sulphur dioxide levels rose by over 500 per cent above normal and persisted for four days. More than 4 000 people, mostly sufferers from bronchitis and heart complaints, died as a result of this pollution episode.

6. Photochemical smog is produced when the exhaust gases of motor vehicles react photochemically to produce *peroxyacetyl nitrates* (P.A.N.) which are extremely active against living organisms. P.A.N.s are formed when a high concentration of motor exhaust gases occurs in bright sunlight and when the air temperature is above about 20°C. P.A.N.s are most common in cities such as Los Angeles but have been shown to occur during hot summers in Britain.

7. The effect of smoke on plants is both direct and indirect. The major direct effect is the blocking up of stomata which inhibits gaseous exchange. The most important indirect effect is the blocking out of sunlight by smoke and smog (smoke particles in water droplets and not to be confused with photochemical smog). Smoke pollution is worst in winter when there is only limited plant growth, but it can have an effect in autumn and spring. In Britain at least, the decrease in smoke pollution has meant an increased amount of winter sunshine in most cities. The hours of winter sunshine in London, for example, increased by 70 per cent between 1950 and 1970.

8. The effect of sulphur dioxide on plants is to cause permanent damage if the concentration of SO_2 in the atmosphere is above 0·25 parts per million (p.p.m.) for several days or more and to cause reversible damage and loss of growth at lower concentrations. Plants vary greatly in their susceptibility to sulphur dioxide pollution and a list of some susceptible and tolerant plants is as follows:

Susceptible	*Tolerant*
Barley and wheat	Corn (maize)
Most leafy vegetables	Cabbages
Beans	Root crops
Larch	Cedar
Birch	Ash
Most pines	Corsican pine
Strawberries	Privet

Sulphur dioxide enters leaves through stomata and disrupts metabolic processes in mesophyll cells. Very old and young leaves are less affected than mature leaves and these show patches of dead tissue in between the veins. Lichens are much more susceptible to sulphur dioxide than higher plants and very few lichens are found in and around large towns and cities. Fungi are similarly susceptible and some fungal diseases such as *black spot* of roses are uncommon in areas of sulphur dioxide pollution.

9. The effect of photochemical smog, when present in toxic concentrations, is to give leaves a waxy or glassy sheen, caused by the collapse of damaged mesophyll cells and their replacement by air pockets. The cells collapse because the P.A.N.s affect the membranes and make the cells leak. Concentrations of P.A.N.s as low as 15 parts per thousand million are sufficient to damage many susceptible plants such as apricots, grapefruit, walnut, clover, oats and many ornamental plants.

WATER POLLUTION

10. Causes of water pollution. According to the river pollution survey, there are about 1 200 miles of river in England and Wales that are so polluted that the water cannot be used for domestic supplies even after treatment. While this is only about 5 per cent of the rivers surveyed, the heavily polluted rivers are almost invariably the lower reaches of large rivers and involve large masses of water. They also flow through areas where millions of people live.

River pollution is the major problem of pollution of fresh water in Britain as most lakes are in upland areas, away from the most common sources of pollution. In other countries, pollution of major lowland lakes is a very serious problem. Pollution of rivers has three main causes; pollution by non-toxic organic wastes, pollution by toxic chemicals, and thermal pollution by hot water.

(*a*) *Non-toxic organic wastes* originate mainly as effluent from sewage works, but also from industry and from farms. The organic matter encourages the growth of micro-organisms which feed on it and break it down. In so doing, however, the micro-organisms can use up the oxygen in the river and lower the oxygen level so much that many other organisms, especially fish, die through lack of oxygen. This is because water contains only 3 per cent as much oxygen as air, even under optimum conditions. Thus fish and other organisms are killed indirectly from the pollution rather than being poisoned by it. The organisms depleting the oxygen supplies are mainly bacteria and fungi. Higher plants are not always harmed because of their low requirement for oxygen, but waste products of the micro-organisms can affect them.

(*b*) *Toxic chemicals* in rivers can affect all forms of life, plant as well as animal, and can lead to the total destruction of life. The pollutants most commonly result from industrial activity and while less common than pollution from organic wastes, the results can be far more disastrous and it can take a river system many months or years to return to normal.

(*c*) *Thermal pollution*, the pollution of rivers by hot water, has a similar effect to that of organic matter. The reason is that oxygen levels decrease in water with increases in temperature. However, some species of plants (and animals) will grow in warm water and thermal pollution may not necessarily destroy the life in a river, only change it to rather different forms.

11. Eutrophication. An effect of pollution particularly relevant to the study of plants is eutrophication. A normal upland river is known as *oligotrophic*, meaning that it is poor in nutrients. Plants do not, therefore, grow very well, the river remains clear and animal life such as fish tend to be dependent on food sources such as insects from outside the river. When such a river reaches lowland areas, it is likely to accumulate mineral salts washed off rich land and eventually becomes *eutrophic*, or rich in nutrients, supporting a much more luxuriant growth of algae and other plants and no longer remaining clear.

Eutrophication also occurs when, through man's activities such as excessive use of fertilisers, the mineral salt content of a river or lake rises unexpectedly. In extreme cases this leads to massive growth of green algae and an *algal bloom*. Eutrophication can deoxygenate a river because, while the algae themselves produce oxygen, this is only in the surface layers. Deeper water is shaded, thus preventing plant growth, and the decomposing remains of dead algae are food for bacteria and fungi which deplete oxygen levels as they do in cases of organic waste pollution.

LAND DERELICTION

12. Land dereliction, like air and water pollution, is a major environmental problem in industrialised countries but, whereas pollution is largely a result of current activity, land dereliction can often be a relic of past industry. There are three main forms of land dereliction.

(*a*) *Excavations*. These are frequently least damaging in terms of visual impact. They usually result from open-pit (i.e. open-cast) mining for minerals, clay, gravel, coal and other physical resources.

(*b*) *Spoil heaps*. Usually the most unsightly forms of land dereliction, spoil heaps can result either from spoil material from excavations or underground workings, or else from the waste products of many industries. If the spoil material is not directly toxic to plants, then it may be recolonised naturally. If toxic, as in the case of tailings from copper mining, the tips may remain bare for decades.

(*c*) *Ruins*. The remains of old factories, mineral extraction plant, mining machinery and buildings and even abandoned buildings can result from the decline of an industry as a result of depression or perhaps exhaustion of a mineral or energy resource.

13. Land dereliction in Britain. Official estimates suggest that there are 55 000 ha of derelict land in Britain. Of this, approximately 40 per cent consists of spoil heaps, 34 per cent of excavations, and 26 per cent of ruins and other forms of dereliction. Land dereliction is most common in particular parts of Britain. Thus South Wales, Lancashire, Yorkshire, Nottinghamshire and Durham have large areas of spoil heaps, mainly as a result of coal mining. The West Midlands has spoil heaps and pits in the Potteries and Cornwall has similar dereliction resulting from china clay extraction. South-eastern England has far less land dereliction than most other parts of the country and most of that consists of excavations resulting from gravel and brick-clay workings.

14. Land reclamation—excavations. Pits resulting from gravel workings or flashes (flooded depressions) resulting from subsidence of underground workings can revegetate naturally to produce areas of high amenity value or even with a potential for wildlife conservation. Reclamation usually consists of encouraging natural revegetation by, for example, the creation of shallow slopes to encourage reed growth, and also planting trees around the borders of excavations. Dry pits are most often filled with waste material and covered with topsoil before seeding with grass for subsequent agricultural use or planting with trees.

NOTE: Fairburn Ings, 20 km south-east of Leeds in West Yorkshire, is a 300 ha nature reserve formed by the flooding of depressions caused by mining subsidence. Now a nature reserve of the Royal Society for the Protection of Birds as well as being a Regional Wildfowl Reserve, it is an internationally known refuge for migrant and wintering wildfowl.

15. Land reclamation—spoil heaps. The two central problems in revegetating spoil heaps, provided the material is not toxic to plants, are counteracting natural infertility and enabling plants to colonise steep, unstable slopes. The former may be overcome by use of fertilisers and possibly lime if the spoil is acid. Steep slopes can be *graded* to make them more shallow, or they can be terraced, or the plants can be spread over the slopes by *hydroseeding*. This process involves spraying a liquid slurry containing seed, fertilisers and a temporary binding agent over the slopes to stabilise them until plants can colonise the spoil heap.

Spoil heaps containing toxic material may be very expensive to reclaim, the toxic material having to be neutralised and top-soil brought in before planting can commence. Where

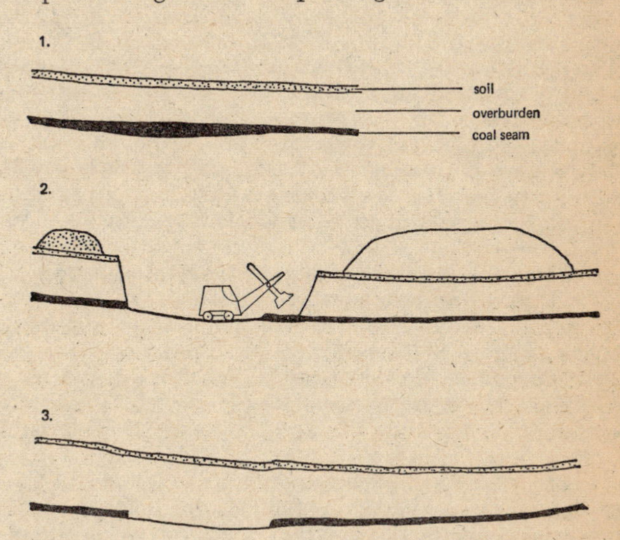

Fig. 60.—Open-cast mining. 1. before mining. 2. mine in operation. 3. mine closed, top-soil and overburden returned.

the toxicity is due to heavy metals such as copper, as in the dereliction around old copper workings, it may be possible to plant special varieties of grasses selected because of their ability to survive in conditions of heavy metal contamination.

16. Open-cast mining is commonly used in Britain for extracting coal and iron ore and involves using large excavators to strip away top-soil and *overburden* to get at mineral or coal deposits up to 100 metres below ground level. In this kind of mining the land can be returned to its previous use very easily and Fig. 60 shows the process, with the whole sequence taking as little as 3 years.

Once the overburden and top-soil has been replaced, the land may be of a higher quality than before. Land suitable for agriculture can, for example, be improved by the installation of field drainage during reclamation. One disadvantage of open-cast mining, however, is the loss of mature trees.

CONSERVATION

17. The need for conservation. In a heavily populated country such as Britain there are very few areas with a natural vegetation cover, that is vegetation not affected in some way by man's activities. Even at the global level, most habitats have been affected by man and the scale of effects continues to increase. Problems such as land, water and air pollution, extinction of animal and plant species, overgrazing of land, destruction of forests and the spread of deserts not only affects the natural environment but also decreases the availability of resources. *Conservation* is an attempt to allow human activity to continue but without causing permanent damage to ecosystems. To conserve an ecosystem is to manage it so that man can benefit from it without destroying the processes essential to its survival. Conservation therefore involves thinking and planning for the long term.

18. Conservation of plant communities in Britain. Although human activity has dominated the British landscape, there are large areas of relatively wide countryside remaining and many semi-natural plant communities survive. A major aim of nature conservation in Britain is to allow representative areas of different kinds of habitat to survive and evolve. Apart from the amenity value of such conservation, it is essential if the study of the biology and especially the ecology of natural communities is to develop in the future. In the past half century various methods of achieving nature conservation have been attempted including national parks, local nature reserves, and work by many voluntary organisations.

19. National parks. There are ten national parks in England and Wales: the Brecon Beacons, Dartmoor, Exmoor, the Lake District, the North York Moors, Northumberland, the Peak

District, the Pembrokeshire Coast, Snowdonia and the Yorkshire Dales. They were set up by the National Parks Commission, established by Act of Parliament in 1949 and now incorporated into the Countryside Commission. All the national parks are relatively sparsely populated, with varied scenery and a diversity of wildlife. The park areas have stringent planning controls compared with other rural areas, but there is not necessarily any particular effort to preserve plant habitats.

20. Nature reserves. Protection and conservation of plant communities is more frequently achieved through the establishment of nature reserves. These may be privately owned or else run by a voluntary organisation such as the National Trust, but there are also a number of *national nature reserves* run by the Nature Conservancy Council. Many counties have *county naturalists' trusts* which run *local nature reserves* often affording some of the best conservation of rare or endangered habitats. In addition, some protection is afforded to an area of land if it is declared to be a *site of special scientific interest* (S.S.S.I.).

21. Voluntary organisations. These have been mentioned in the previous section, but several national organisations have become important in the conservation of plant communities. The Botanical Society of the British Isles includes both amateur and professional botanists among its members and collaborates with research centres in promoting conservation of plant communities. The National Trust was initially concerned with preserving famous old buildings but has now grown to the point where it controls over 1 000 properties and over 170 000 ha of land, including some of the finest areas of natural and semi-natural vegetation left in Britain. The Field Studies Council is, on the other hand, primarily an educational body, but in a very practical sense in that it runs a network of field centres throughout England and Wales which run courses on all aspects of field studies, including conservation.

22. Research. Effective conservation of plant communities is greatly enhanced by continued research. Much of this is carried out in universities and colleges, often with finances provided by the Natural Environment Research Council.

This also finances the Institute of Terrestrial Ecology and other research stations. The Institute was formerly the research wing of the Nature Conservancy and includes two world famous research stations, Monkswood near Huntingdon and Merlewood in Cumbria.

There are many organisations and journals through which ecologists communicate the results of their research and their ideas. The biggest in Britain is the British Ecological Society which publishes three journals of international repute, the *Journal of Ecology*, the *Journal of Animal Ecology* and the *Journal of Applied Ecology*.

PROGRESS TEST 12

1. What is a pollutant? (1)
2. What are the most important kinds of air pollution? (2)
3. What are the effects of sulphur dioxide on plants? (8)
4. List six plants that are susceptible to sulphur dioxide pollution. (8)
5. How does photochemical smog affect plants? (9)
6. What are the major causes of water pollution? (10)
7. What is eutrophication? (11)
8. What are the main forms of land dereliction? (12)
9. What are some of the methods used to reclaim spoil heaps? (15)
10. Describe some of the ways in which plant communities are conserved in Britain. (18–20)

APPENDIX I

GLOSSARY OF TERMS

Abscission layer: a layer of weak cells produced in a leaf base prior to leaf fall.

actinomorphic: capable of vertical bisection in two or more planes to give identical halves.

active transport: that form of movement of substances within a plant that requires an input of energy.

aecidium: a structure producing asexual aecidiospores in some basidiomycetes.

aerobic respiration: respiration requiring oxygen.

alternation of generations: the alternation, in the same species, of haploid (gametophyte) and diploid (sporophyte) plants.

amyloplast: a plastid containing starch.

anaerobic respiration: respiration which can proceed in the absence of oxygen.

anaphase: the fourth phase of mitosis in which the chromosomes divide into free chromatids which then separate.

androecium: the cluster of stamens on a flowering plant.

annulus: a ring of cells attaching the operculum to the capsule in some lower plants.

anther: the microsporangium at the top of a stamen in flowering plants.

antheridium: that part of a non-vascular plant that produces male gametes.

antherozoid: a small motile male gamete produced by some lower plants.

antibiotic: a substance produced by some lower plants which can prevent the growth of other organisms, especially those causing disease.

archegonium: female sex organ of liverworts, mosses, ferns and most gymnosperms.

ascospore: a spore produced by an ascomycete fungus.

ascus: the spore-producing body in an ascomycete fungus.

asexual reproduction: reproduction which does not involve the combining of two gametes.

autecology: the ecological study of a single species.

autotroph: an organism which makes its own food using an external energy source, such as sunlight, and external sources of organic substances.

auxin: a kind of plant hormone which may control secondary growth, abscission, tropisms and other plant functions.

back-cross: the crossing of a hybrid plant with a plant genetically identical to one of its parents.

bacteriophage: a virus that infects bacteria.

basidiocarp: the fruiting body, e.g. mushroom, of a basidiomycete fungus.

basidiospore: the spore produced by a basidiomycete fungus.

basidium: the spore-producing structure in a basidiomycete fungus.

biogeochemical cycle: the process by which an element is cycled in different forms and which ensures its continued availability to living organisms.

bulb: a storage and propagative organ, formed, usually below ground level, from a swollen bud.

Calvin cycle: a cyclic process of chemical reactions occurring in photosynthesis.

calyptra: a hood-like covering of part of the sporophyte in bryophytes and of the fruits in some higher plants.

calyx: a ring of sepals on a flowering plant.

capsule: a compact slime layer surrounding some kinds of bacteria; another term for the theca on the bryophytes.

carbohydrate: an important kind of chemical compound containing carbon, hydrogen and oxygen with a 2:1 hydrogen:oxygen ratio.

carpel: the female organ of a flowering plant.

carpospore: a flask-like haploid female structure found in red algae, i.e. *Rhodophyceae*, etc.

cell membrane: the membrane which surrounds a plant cell, inside the cell wall.

cell plate: the first stage in the formation of a new cell wall when a cell divides during the telophase stage of mitosis.

cell wall: the semi-rigid structure made up largely of cellulose, a carbohydrate, which encloses a plant cell.

centromere: the point on a chromosome at which the two chromatids are joined together.

chemoautotroph: an autotroph which uses chemical energy.

chlorophyll: a green pigment essential for photosynthesis.

chloroplast: a plastid containing chlorophyll which is involved in photosynthesis.

chromatid: one of two spirally arranged filaments which makes up a chromosome.

chromatin: the dispersed material found in the nucleus which, during cell division, forms chromosomes.

chromoplast: a coloured plastid common in petals.

chromosome: a long thin body which contains genetic information and is found in the nucleus of a cell.

codon: that portion of a DNA molecule, called a triplet, which codes for a particular amino acid.
coenocyte: a multinucleate hypha or filament without cross walls.
collenchyma: thick-walled vacuolated cells which give structural support in higher plants.
community: all the living organisms in an ecosystem.
companion cell: a small cell with dense cytoplasm occurring alongside a sieve tube.
conceptacle: flask-shaped vessel producing gametes in some algae.
conidiophore: a fungal hyphum, often aerial, producing conidia.
conidium: an asexual spore produced by some kinds of fungi.
contractile vacuole: a contractile sac in a cell which helps to control water and salt content.
corm: a storage organ consisting of a thickened stem base with thin scale leaves.
corolla: a ring of petals.
cortex: a core or ring of tissue in a plant, often with structural and conductive functions.
cotyledon: the first leaf to develop in a young seed plant, often fleshy and constituting a food store.
cristae: infoldings of the inner of two membranes in a mitochondrion.
cuticle: the layer of cutin and other compounds covering the surface of a plant.
cutin: the waxy waterproof substance commonly produced by the cells in the surface layers of a plant.
cytoplasm: a jelly-like material found in living plant cells.
deme: a group of organisms which interbreeds.
dicotyledon: a plant producing two cotyledons.
dictyosome: a stack of smooth-membrane sacs in the cytoplasm of a cell which is involved in transport and secretion.
dictyostele: a network of tubular vascular tissue in ferns.
dioecious: a plant producing male or female gametes, but not both.
diploid: an organism with two sets of chromosomes, one derived from each parent.
dominant: a gene or its character which manifests itself in a heterozygous individual.
ecosystems: a defined area on the earth's surface including all the living organisms and physical factors in that area.
edaphic factor: a factor concerned with soil, e.g. micronutrient availability.
embryo sac: the angiosperm female gametophyte, usually an eight-celled structure.
endergonic reaction: a chemical reaction which requires energy to proceed.

endodermis: a water-resistent layer of cells enclosing a root stele.
endoplasmic reticulum: a complex network of membranes within the cytoplasm, often with ribosomes embedded in it.
endospore: a thick-walled resistant resting spore produced by some bacteria.
enzyme: a biological catalyst which enables a chemical reaction to proceed much faster than would otherwise be the case.
epidermis: the outermost layer of cells of a plant.
eutrophy: an excess of organic matter in a body of water.
excretion: the removal of waste substances from an organism.
exergonic reaction: a chemical reaction that liberates energy.
fibre: a long thin tough cell found in sclerenchyma.
filament: the slender flexible stalk of a stamen.
flagellum: a whip-like appendage of a cell which enables an organism to swim.
fruit: a seed enclosed in modified parts of the ovary.
gamete: a sex cell.
gametophyte: a haploid plant body which produces gametes.
gemma: a ball of cells produced by some bryophytes in asexual reproduction.
gene: the portion of a chromosome which controls a specific characteristic of an organism, e.g. seed colour.
gene frequency: the frequency with which a gene occurs compared with the other types of gene affecting the same character.
gene pool: all the genes found within a deme.
giberellin: one of over thirty closely related hormones which can control plant growth.
glycolysis: the breakdown of glucose; part of the respiratory process.
golgi body: a dictyosome.
grana: stacks of flattened membranous sacs in chloroplasts, having the appearance of piles of plates.
gynaecium: a cluster of carpels.
halophyte: a plant tolerant of very saline (salty) conditions.
haploid: a cell with only one set of chromosomes.
haustorium: a pad of tissue produced by a parasitic plant which aids penetration of a host's tissues.
hearth of domestication: a site where agriculture first developed.
hermaphrodite: bisexual, having stamens and carpels on the same flower.
heterogamy: the condition of having male and female gametes of different sizes.
heterosporous: an organism producing more than one kind of spore.
heterotroph: an organism which cannot make its own food and therefore consumes other organisms, their products or remains.
heterozygous: a condition where a pair of genes are not alike and

an organism does not breed true for the character determined by those genes.

homozygous: a condition where a pair of genes are alike and the organism breeds true for the character determined by those genes.

hydrophyte: a plant adapted to very wet or even submerged conditions.

hypha: the thin thread of cells of which fungi are composed.

indusium: a membraneous covering over a sorus.

interferon: a protein substance produced by many animals which is harmful to viruses.

interphase: the resting stage between each division of a cell.

involucre: a flap of tissue protecting an archegonium.

irritability: the ability of an organism to sense and respond to changes in its environment.

isogamy: the condition of having male and female gametes of the same size.

kinin: a hormone which controls the rate of cell division.

Krebs cycle: a cycle of chemical reactions central to aerobic respiration.

lamina: the flattened blade of a leaf or leaf-like organ.

lenticel: a channel or pore of cork-filled cells in the stem of higher plants.

lethal gene: one which kills the organism when it expresses itself.

lignin: a chemical constituent of cell walls.

ligule: a membraneous outgrowth of a leaf.

lipid: a chemical made up of glycerol and fatty acids and important in maintaining the structure of a plant and in storing energy.

lysosome: a cell organelle comprising a small sac containing enzymes.

macronutrients: the six elements, other than carbon, hydrogen and oxygen, required by plants in large amounts: nitrogen, potassium, phosphorus, calcium, sulphur and magnesium.

medulla: the pith or central core of a higher plant stem; the central tissue of the thallus of an alga.

medullary ray: the tissue extending from the pith to the cortex between vascular bundles in flowering plants.

megaspore: the larger of two kinds of haploid spores produced by a plant.

megasporophyll: a modified leaf bearing mega-, or macro-, spores.

meiosis: a form of cell division involving a reduction in the chromosome number to half the normal.

membrane: a thin pliable sheet-like structure made up of lipid and protein.

meristem: a region of active cell division in a plant.

mesophyll: photosynthetic parenchyma tissue common in leaves.

mesophyte: a plant which is not adapted to survive too much or too little water.
metaphase: the third stage of mitosis and meiosis when chromosomes become attached to the spindle and form a layer at the widest part, or equator, of the spindle.
microfibril: a strand of cellulose, the basic constituent of cell walls.
micronutrient: an essential element required by a plant in small amounts.
micropyle: a pore on an ovule located between the integuments.
microspore: the smaller of two kinds of spores produced by a plant.
microsporophyll: a modified leaf-like structure producing microspores.
middle lamella: the pliable central portion of a cell wall.
mitochondrion: a small cell organelle with a double membrane which is involved in respiration.
mitosis: division of a nucleus involving duplication and separation of chromosomes.
monocotyledon: a plant with a single cotyledon.
monoecious: a plant producing male and female gametes.
monohybrid cross: a cross between plants differing in only one character.
mutagen: a chemical capable of inducing mutations.
mutation: a spontaneous change in form or function of an individual caused by a change in gene structure.
mycelium: a tangled mass of hyphae making up the main body of a fungus.
mycorrhiza: fungi closely associated with the roots of some higher plants, especially orchids and conifers.
myoneme: a contractile fibre found in some algae.
nucellus: the mass of cells from which an ovule develops.
nucleic acid: a compound made up of sugars, phosphoric acid and nitrogen-containing compounds; very important in heredity and in controlling cell function.
nucleolus: a dark-staining portion of the nucleus containing ribose nucleic acid.
nucleus: the part of a cell controlling cell function and reproduction and containing the chromosomes.
oligotrophy: a body of water deficient in mineral elements and hence in organic matter.
oogamy: the condition of having a large non-motile female gamete and a small male gamete.
oogonium: a structure producing an oosphere in algae and fungi.
oosphere: a large non-motile female gamete in algae and fungi.
oospore: a resting spore produced in some lower plants after fusion of male and female gametes.
operculum: a circular lid on top of a theca.

osmosis: the passage or diffusion of small molecules across a membrane impermeable to larger molecules.

ostiole: a pore or hole in a conceptacle; found in some algae and fungi.

ovary: an ovule protected by additional tissue, as in angiosperms.

ovule: the megaspore retained within a megasporangium and covered with a protective integument; forms the seed after fertilization.

parabiosphere: part of the biosphere where life exists only in dormant forms.

paraphysis: a sterile hair found in sex organs in some simple plants.

parasite: a heterotroph which lives off living organisms.

parenchyma: large structural cells with photosynthetic, food storage and/or structural functions.

pellicle: the elastic coat covering some simple algae.

perianth: the petals and sepals of a flowering plant.

peristome: a ring of "teeth" in a capsule which aids spore dispersal.

permanent wilting point: the degree of wilting beyond which a plant cannot recover.

petal: a modified, and often coloured, leaf surrounding stamens and carpels.

phloem: conductive tissue responsible for translocation.

phospholipid: a lipid containing phosphate groups.

photoautotroph: an autotroph which uses light as its energy source.

photosynthesis: synthesis of complex foods from simple chemicals using the energy of sunlight.

phytochrome: a hormone affected by light; it may control flowering and seed germination.

phytohormone: a plant hormone.

pinna: the leafy part of a fern leaf.

pinocytosis: a process by which insoluble material or fluid is surrounded by the cell membrane prior to being engulfed or excreted by the cell.

plasmalemma: the cell membrane.

plasmodesmata: thin strands of cytoplasm which connect the contents of one cell to those of a neighbour through a pore in the cell wall.

plastid: a general term for a number of ovoid bodies of varying function found in plant cells, e.g. chloroplast.

plumule: the initial shoot produced within a seed.

polyploid: a plant or cell in which the number of chromosomes is more than twice the haploid number.

polysome: a chain of ribosomes embedded in the endoplasmic reticulum.

population: the members of a species within a particular community.

primary producer: an organism which makes its own food.
prometaphase: the second stage of mitosis when the nuclear membrane ruptures and a spindle develops.
prophase: the first stage of mitosis and meiosis, when chromatin coalesces into chromosomes.
protein: a chemical made up of nitrogen-containing amino acids; the major constituent of enzymes and many other important components of living organisms.
prothallus: a fern gametophyte.
protonema: a small branched structure in mosses developing from spores.
provirus: the infectious part of a bacterial virus which attaches itself to its host's nucleic acid and divides at the same time as the host.
psychrophile: a cold-tolerant plant.
pyrenoid: a proteinaceous body in the chloroplasts of some algae, which stores starch.
quantosome: a raised area on a granum of a chloroplast, containing chlorophyll.
rachis: the stiff rod-like part of a fern leaf.
radicle: the initial root produced within a seed.
receptacle: the fertile end of an algal frond; the tissue of a flowering plant bearing stamens and carpels.
recessive: a gene or character which does not manifest itself in a heterozygous individual.
respiration: the chemical breakdown of food to release energy.
rhizome: an underground stem, usually thick and fleshy.
ribosome: a small body containing RNA and protein and involved in protein synthesis.
runner: a horizontal stem which can bud and root to produce a new plant.
saccharide: the basic unit of a sugar.
saprophyte: a heterotroph which consumes the dead remains of other organisms.
sclerenchyma: thick-walled lignified supporting cells, usually dead.
seed: the mature ovule of a flowering plant with a protective covering and often a food store or endosperm.
sepal: a tough leaf-like appendage protecting a young flower.
seta: the stalk of a moss sporophyte.
sexual reproduction: production of an organism by the fusion of male and female gametes.
sieve tube: a long thin non-nucleated but living cell found in phloem.
sorus: an organ in ferns containing sporangia.
spermatium: a non-motile male gamete produced in some fungi and algae.
sporangium: a structure producing spores.

sporophyte: a diploid plant body producing haploid spores.
stamen: the male organ (microsporophyll) of a flowering plant.
stele: the core of conducting tissue in a root.
stigma: a receptive surface for pollen grains at the end of a style.
stolon: a slender underground stem.
stomata: pore-like openings in the leaves of higher plants.
stomium: a pore in a fern sporangium from which spores are released.
strobilus: a reproductive structure in a horsetail.
stroma: the fluid contents of a chloroplast.
style: a slender structure bearing a stigma.
suberin: a waxy waterproof substance common in corky tissue.
symbiosis: when two organisms of different species live together with benefit to each.
synecology: the ecological study of relations between species.
eleutospore: an asexual spore produced by some basidiomycetes.
elophase: the final stage of mitosis and meiosis when each set of chromatids or daughter nuclei forms a new nucleus.
esta: a hard coat surrounding the embryo and endosperm in a seed.
heca: the spore-producing capsule of a moss sporophyte.
hermophile: a heat-tolerant plant.
onoplast: the membrane lining a vacuole.
racheid: a very long thin dead cell with pitted walls found in xylem.
ranslocation: movement of photosynthetic products and other complex chemicals around a plant.
ranspiration: the process by which plants lose water from the leaves and generate an upward flow of water from the roots.
ropism: a response by a plant to an external stimulus such as light or gravity.
ube cell: part of a pollen grain.
urgor pressure: pressure in a cell resulting from the cytoplasm taking in water.
uredospore: an asexual spore produced by some basidiomycetes.
acuole: a sac in the cytoplasm of a cell, containing sap.
ariation: variability of form or function within a deme.
ascular bundle: a bundle of conductive tissue.
ector: an organism which carries a parasite from one host to another.
egetative reproduction: reproduction by modified plant parts, e.g. bulbs and corms, without sexual processes; also known as asexual reproduction.
essel: A very long thin dead cell without end walls, found in xylem.
irus: a very small simple parasitic organism.
xerophyte: a plant adapted to live in dry conditions.

xylem: a structurally strong tissue which conducts water and mineral salts from roots to leaves.

zoospore: a motile spore produced, in particular, by many algae and fungi.

zygomorphic: bilaterally symmetrical.

zygote: a diploid cell resulting from the fusion of two gametes.

APPENDIX II

EXAMINATION TECHNIQUE

To ensure success in an examination, a candidate should: know and understand the aspects of the subject covered by the syllabus; demonstrate this understanding under examination conditions. Without the former, no amount of hints and suggestions will ensure success, but provided the candidate has learnt and understood the subject, then the following notes on examination technique may be helpful.

1. The changing nature of examinations. Until fairly recently most G.C.E. examinations in biological subjects consisted largely of fairly general questions. Each paper gave a candidate a choice of perhaps four or five questions, to be selected from a total of seven to ten. This meant that a candidate could (and often did!) pass an examination without being familiar with all of the syllabus.

With may examining boards it is now more usual for an examination to include a large number of compulsory questions, each question being in several parts but with each part requiring only a short answer. Such questions are usually complemented by more general questions, so that a candidate now has to demonstrate a general knowledge of the whole syllabus together with a more detailed knowledge of a substantial part of it. Examples of multiple-part questions requiring short answers are given in Appendix III, but the reader is referred to the Progress Tests at the end of each chapter for further assistance.

2. The marking scheme. Most examiners are experienced teachers and although a group of examiners concerned with a particular paper will have a marking scheme to aid them, this is likely to be very brief. It will often be concerned largely with apportioning marks for different parts of a question. The student should therefore learn to estimate the relative importance of parts of a question and this is much easier if one is familiar with the syllabus and has looked at previous papers. A two-part question may involve a simple definition followed by two examples of the process or phenomenon defined. For example:

"What is a biogeochemical cycle? Give an account of two such cycles."

The first part would be allocated 6 marks out of 20 and the second part 14 marks, 7 for each cycle described.

3. Timing. Five questions in 2½ hours means 30 minutes per question but in practice one should spend the first 10 minutes reading the examination paper and deciding which questions to answer and the last 5 minutes checking answers for obvious errors, especially in diagrams. Answer the questions about which you feel most confident first, but never overrun the time allocation by more than a couple of minutes. Even an "impossible" question left until last can often be answered, especially if there is time enough to give it some serious thought.

4. Diagrams and sketches. It is usually possible to present information quicker in a diagram or table than in writing. In some circumstances lack of a diagram may lose marks. If time allows, the underlining of key words will be appreciated by a hard-pressed examiner.

5. Relevance. "Answer the question" is a frequent but important plea. Irrelevant "padding" does not earn marks and the time is better spent on other questions or in endeavouring to construct a short but relevant answer. The longest answer does not necessarily earn the best marks.

6. Answer the stated number of questions. You lose 20 per cent of the marks by not attempting the fifth question. The same principle applies with compulsory questions where even an educated guess is better than no answer at all. Incidentally, it is better to attempt all the questions, even if one answer is below standard, than to concentrate on fewer good answers. It is usually easier to pick up 4 marks at the beginning of an answer than to spend the time trying to improve an answer already worth perhaps 14 marks out of 20.

7. Use the space available. While no-one should waste paper, it is pointless to draw tiny diagrams or graphs. Large, neat diagrams greatly enhance an answer.

8. Problems. A student of reasonable all-round ability will usually score marks more easily for "problem" type questions with definite answers than for "essay" type questions.

9. Never leave before the end. Many a student has left an examination hall early, only to find his memory improving with unpleasant rapidity! It is better to spend spare time checking answers or even re-drawing diagrams. Often, further facts and ideas will present themselves.

APPENDIX III

QUESTIONS

EXAMPLES OF COMPULSORY MULTI-PART QUESTIONS

1. (*a*) By means of labelled diagrams show the tissues external to the xylem in the stem of a tree in

(*i*) a sector of a transverse section of a stem before secondary thickening

(*ii*) a corresponding sector of a five year old tree.

(*b*) In girdling the shoot of a tree, a ring of bark (all tissues external to the xylem) was removed from the entire circumference. After several years the shoot died.

(*i*) Why did the shoot not die immediately?

(*ii*) Why did the shoot eventually die?

(*iii*) What difference is likely to arise between the vegetative growth of the shoot above and below the girdle? Explain your answer.

(*iv*) How does the girdling of branches of fruit trees result in the production of larger fruit?

(*J.M.B. Biology A*)

2. (*a*) What is understood by the term heterotrophic nutrition?

(*b*) What structural and physiological features of a named fungus may be related to this form of nutrition?

(*c*) By reference to examples, give three different ways in which fungi are

(*i*) beneficial to man.

(*ii*) harmful to man.

(*J.M.B. Biology A*)

3. How does mitosis maintain, and meiosis lead to variation in, the genetical constitution of organisms?

At what stage in the life of the following does meiosis occur?

(*i*) A *named* alga?

(*ii*) A liverwort, a moss or a fern?

(*iii*) A flowering plant?

(*J.M.B. Biology A, modified*)

4. Why is the sun essential for the continuing existence of life on Earth?

(*A.E.B. Environmental Studies O*)

5. State two ways in which fungi differ from green plants.

(*A.E.B. Environmental Studies O*)

6. Explain why each of the following questions is technically incorrect.

(*a*) All fungi are parasites.
(*b*) Seeds are produced only by flowering plants.
(*c*) Mosses and ferns do not depend on water during sexual reproduction.

(*A.E.B. Biology O*)

7. Give concise accounts of the form, general location and fine structure of any *four* of the cell components listed below. State, briefly, the function or functions performed in the cell by each of the components you have chosen.

(*i*) cell membrane (plasmalemma).
(*ii*) interphase nucleus.
(*iii*) endoplasmic reticulum.
(*iv*) plant vacuole.
(*v*) contractile vacuole.
(*vi*) lysosome.
(*vii*) Golgi apparatus.

(*J.M.B. Biology A, modified*)

EXAMPLES OF GENERAL QUESTIONS

8. By reference to *named* plants describe, giving details, four different ways of increasing plant production.

(*A.E.B. Biology O*)

9. Explain simply how genetic information is replicated during the mitotic division of a chromosome. How may this information be conveyed to the cytoplasm and thereby control cell processes?

(*J.M.B. Biology A*)

10. Explain what is meant by "alternation of generations" by reference to the life histories of a moss and a fern.

(*London Biology A*)

11. Relate the anatomy and morphology of the leaf of a terrestrial flowering plant to its functions. Bearing in mind the requirements of a terrestrial leaf, what modifications would you expect to find in a floating or submerged leaf of a water plant?

(*J.M.B. Biology A*)

12. Describe the primary root of a herbaceous plant as seen in a median longitudinal section. Describe the changes which

occur in the size and structure of the cells with increasing distance from the tip.

(*J.M.B. Biology A*)

13. What do you consider to be the mechanism of evolution? Give briefly the evidence to support your arguments.

(*O. and C. Biology A*)

14. Describe a generalised plant cell as seen under the light microscope. What further features may be seen with an electron microscope?

15. Write an essay on viruses.

16. What is a lichen? Give an account of the economic and ecological significance of lichens.

17. Describe the structure and process of sexual reproduction of a named brown alga. What are the general features of sexual reproduction in red algae which differ from brown algae?

18. List the functions of four macronutrients in higher plants. How are deficiencies of these macronutrients expressed as symptoms in plants? What are the two processes by which such macronutrients are taken up by plants?

19. Describe, in outline only, the light and dark reactions of photosynthesis. List three factors which affect the rate of photosynthesis.

20. Give an account of the physical factors which help to determine the kinds of plants to be found in a particular area.

21. Describe the major ways in which crop production may be improved. What are the causes of world food shortages?

22. Give an account of the causes and extent of the major forms of air pollution and describe their effects on plants.

23. What are the three major forms of land dereliction in Britain and most industrialised countries? Indicate some of the ways in which derelict land may be reclaimed.

24. Describe three cereal crops. Give a brief account of three important *tropical* fruit crops.

J.M.B.	Joint Matriculation Board.
A.E.B.	Associated Examination Board.
London	University of London University Entrance and School Examinations Council.
O. and C.	Oxford and Cambridge Schools Examination Board.

INDEX

Basic Biology

P. T. MARSHALL

The main theme of this HANDBOOK is the elementary physiology of the green plant and the mammal. "All that is likely to be needed for "O" Level or C.S.E. Biology is included, from the cell and the chemistry of important compounds to nutrition – both plant and mammal, respiration, co-ordination, internal control and reproduction. In short, a book well described by its title, and useful to teacher and pupil." *The Teacher*
Illustrated

Biology – Advanced Level

P. T. MARSHALL

This HANDBOOK is compiled along the lines of the new and fully revised syllabuses in Advanced Level biology that are now operating in a number of boards. The text has been completely updated for the latest edition particularly with regard to genetics and plant co-ordination.
Illustrated

Human and Social Biology

GEORGE USHER

This HANDBOOK is based on the newly revised "O" Level syllabuses now operating in a number of boards. Relevant sections have been expanded as necessary to include the material required for the Health Science syllabuses used by many overseas candidates.
Illustrated

Genetics

M. W. ROBERTS

This HANDBOOK has been written for the purpose of providing a concise treatment of genetics for the student studying for "O", "A" and "S" Level exams. As thorough a framework as is possible is provided by the book, including a simple treatment of the chemistry involved in genetics. Solved problems and a selection of sample exam questions are also included.
Illustrated